Popular Modernity in America

SUNY series in Postmodern Culture
Joseph Natoli, Editor

Popular Modernity in America

Experience, Technology, Mythohistory

Michael Thomas Carroll

STATE UNIVERSITY OF NEW YORK PRESS

Published by
State University of New York Press
© 2000 State University of New York
All rights reserved
Printed in the United States of America

For information, address the State University of New York Press,
State University Plaza, Albany, NY 12246

Marketing by Dana E. Yanulavich • Production by Bernadine Dawes

Library of Congress Cataloging-in-Publication Data

Carroll, Michael Thomas, 1954–
    Popular modernity in America : experience, technology, mythohistory / Michael
Thomas Carroll.
       p. cm.—(SUNY series in postmodern culture)
    Includes bibliographical references (p. 203)
    ISBN 0-7914-4713-8 (hc)—ISBN 0-7914-4714-6 (pb)
       1. Popular culture—United States—History. 2. Modernism (Aesthetics)—United States.
3. Technology—Social aspects—United States. 4. Arts, American. 5. Arts, Modern—20th
century—United States. I. Title. II. Series.

E169.1.C278 2000
306'.0973—dc21
                                          00-020429

1 2 3 4 5 6 7 8 9 10

For my parents,
James Donald Carroll and Catherine Veronica Krimmel

Contents

Acknowledgments

I would like to thank the following colleagues for their encouragement, support, and constructive criticism: Charles Truxillo, Eddie Tafoya, Joseph Slade, Dennis Hall, Roger Rollin, Barbara Risch, Sara Hanna, Gary Burns, Ross Mitchell, David Kaloustian, and Steve Williams. Thanks also to the Faculty Research Committee at New Mexico Highlands University for providing funds for research-related travel, to Patricia Hewitt for research assistance, and to Joan Snider for editorial assistance. I also wish to thank the following organizations for providing me with an audience for the early drafts of these chapters at annual conventions: the Popular Culture Association and the American Culture Association, the American Studies Association, the Philosophy/Interpretation/Culture (PIC) Conference at Binghamton University, and the International Conference for the Fantastic in the Arts. Finally, thanks to the Smithsonian Institution American History Archives and to the Kinsey Institute (and J. P. Yamishiro in particular) for granting me access to their collections.

Introduction

This book is an inquiry into the relationship between technology, culture, and experience in the United States from the 1840s (roughly speaking, the Age of Manifest Destiny) to the 1980s (the Age of Reagan), with a few forays into the 1990s. This is obviously a complex task, but worse than that, it is plagued by conceptual paradox, for the very phrase "cultural experience," considered epistemologically, is oxymoronic. Contemporary cultural studies are generally grounded in the constructivist—and actually, Lockean—assumption that the human subject is a template upon which text is inscribed, and to generate knowledge in the human sciences, one need only reveal the mechanism of that construction. "Experience," on the other hand, suggests something else: according to Husserl, the "crisis" in Western science was precisely the result of our not having come to terms with experience itself, and his phenomenology was based on an attempt to describe the "prescientific" basis of knowledge through the identification of the transcendental ego and innate modes of perception and experience.

In an attempt to negotiate this primary conceptual and procedural difficulty, I have assumed that both textual and experiential approaches are both valid and necessary, and thus my procedural grounding is in keeping with what Don Ihde calls "postphenomenology." Postphenomenology rejects the transcendentalism, foundationalism, and to some extent the methodological formalism of classical (Husserlian) phenomenology's *epochē* (a procedure for reducing phenomena to "essences") while retaining a focus on the experiences of the human subject; these experiences are defined here hypothetically and intersubjectively, but with reference to the way experience is always historically and culturally encoded. In my analysis of the materials, I've found that a probing of cultural experience in the technological age reveals a triadic relationship, the elements of which are mythohistory (the primary vehicle for the ideological and cultural construction of the subject), technology (an intermediary way of knowing and being-in-the-world), and the primary forms of perception/experience; each of them have an active and determining role in the equation. My methodology is loosely structured according to these three elements, while the book's overall organization into chapters reflects the crossing of two expository codes. The first is that of traditional chronology and cultural history (moving from considerations of the primary phase of popular modernity's technologies to their latter deployment in specific entertainment genres); the second is based on an achronological schema related to the phenomenal categories of space, time, sound, and sight, and narrative, which, I will argue, constitutes a unique intersection of mythohistory, technology, and experience.

A key concept in this triadic relationship is hypermediation, which is not to be confused with the terms "hypermedia" and "hypertext," which refer to computer accessing methods. Hypermediation here refers to the way in which subjective experience of the technological is forestructured by media, and it is a useful conceptual tool in that it helps overcome the simplistic culture:experience binary by providing descriptions of the way in which the subject's interactions with the technological matrix of popular modernity is formed by both the parameters of primary human experience and the way in which experience is guided by preestablished mythohistorical and ideological codes.

All the specimens examined here are part of the intertextual matrix that I call popular modernity. While this cultural matrix has historical precursors—i.e., the "bread and circus" society in the city of Rome in the first century A.D., the baroque Catholicism of seventeenth-century Spain (cf. Maraval 1986), the commercial culture exemplified by the late medieval system of trade routes in Europe (cf. Burke 1978), and the public culture constructed in France after the revolution (cf. B. Anderson [1991] 1994; Hobsbawm [1983] 1994), I speak here of a popular culture that began to emerge in the United States in the mid-nineteenth century, a popular culture that combined a mass political ideology (as Gabler says, popular culture was "swept along on the tide of Jacksonian democracy" [1998, 32]), consumerism, and, most importantly for our purposes here, an advanced technological matrix. In addition to this technological definition, I follow Karen Pinkus's use of "popular modernity" to indicate a domain distinct from "modernity in an elite or merely aesthetic sense" (1995, 88), but to which I want to add a relationship between technology and the subject that is different in scope and character from these relationships in both pre- and early-modern epochs. Also important in this regard is David Harvey's notion of Fordism, which refers to forms of social and economic organization that emerged in the first two decades of the century and that explicitly recognized the importance, not only of mass production but of mass consumption as well, along with "a new aesthetics and psychology . . . a new kind of rationalized, modernist, and populist democratic society" (1989, 125–26). Fordism is an important historical midpoint for this book, for if the politics and popular culture of the Age of Manifest Destiny mark the very dawning of popular modernity, then the new cultural and economic doctrine of the 1920s marks its transformation into a doctrine for the management of a mass culture, while the Age of Reagan, as I use the term here, indicates a phase of popular modernity that continues to the present and is characterized by an intensification of popular modernity's dense and totalizing fusion of entertainment, information, technology, and mythohistory. Further, for the most part I follow Bernard Yack's stance toward postmodernism, seeing it as a set of theories and sensibilities rather than as a new historical epoch. Like Yack, my assumption is that the "basic social conditions [of] modernity," a "juggernaut of material change and technological innovation[,] still runs through our age" (1997, 66), and thus, while changes within popular modernity noted above will be indicated, for the most part I will focus on the continuities of popular modernity.

The first chapter of this book broadly canvasses the development of the American notion of space. The discussion begins with a consideration of a very old technology indeed—cartography—and then moves to an examination of the United States during its formation as an industrial nation-state and the role played by the telegraph and the newspapers in the "imagining of the nation," for it is here that the modern concept of American space begins. Also important in this regard is the American concept of frontier as generally derived from Puritan exegetical typology, and, more to our purposes here, Frederick Jackson Turner's "frontier thesis." Turner's pitting of "civilization" against "savagery" is an important axiomatic key to American ideology and experience and an example of the Self-Other dialectic in cultural construction and experience. We then turn to the late 1870s, during which, through the introduction of telephony, the reconfiguration of spatial concepts was brought into more direct contact with mass society. The shift from telegraphy to telephony is an important developmental moment in popular modernity standing roughly halfway between the formation of popular modernity in the 1840s and the ascendancy of Fordism in the 1920s, for while telegraphy was directly experienced by relatively few people, the telephone would become part of daily experience, part of the domestic sphere, and a reshaper of social relations; its use, ultimately, was reshaped by the user. The discussion then turns to postindustrial America for an examination of the continuities and discontinuities of modernity's technospace in the simulacra of television, video games, and cyberspace.

The second chapter continues with a theme introduced in the discussion of telephony by probing the phenomenal nature of the disembodied voice of radio broadcasting. In its first decade (1923–33), broadcast radio contributed to the development of emerging popular modernity, going well beyond telephony in terms of its ability to synergistically combine communications technology, political ideology, myth, and entertainment. The particular qualities of radio broadcasting provide a strange link between the leaders of the world fascist movement in Germany and Italy, the more discreet authoritarianism of Franklin Roosevelt, and the sexual appeal of the pop singer (specifically, the "radio crooner"). As with the first chapter, the experience of media is examined in light of primary engagements between, first, the human subject and technology, and second, the intersubjective framing of media through hypermediation—the way media is represented and looped back to us through other media. The latter is demonstrated in my analysis of one of Mack Sennett's short films, *Sing, Bing, Sing* (1931), featuring the radio crooner par excellence, Bing Crosby. Consideration is also given to the voice of "anonymous authority," a term I've borrowed from Erich Fromm to describe the depersonalized voice of recorded telephone operators.

The third chapter examines temporal experience under the perceptual regime of popular modernity, particularly in regard to entertainment forms that illustrate some of the peculiarities of the modern subject's time orientation; first, the "nostalgia film," from Hollywood musicals of the 1940s to the quintessential pop film of the Reagan era, *Back to the Future* (1985); second, the utopian and dystopian fantasies of science fiction, as well as the fusion of these chronophobic narratives in the peculiar mode of the nostalgic-futorologic, as exemplified by *Star Trek: Generations* (1994); and third, the incessant remarketing of the popular music of the recent past. Here, as elsewhere, my strategy is to examine the dynamic interplay of mythic text, experience and perception, and twentieth-century communications and entertainment technologies, drawing on both the classical phenomenological analysis of time perception and approaches that stress the acculturated aspects of temporality, particularly under the aegis of consumerism.

The fourth chapter is the longest chapter in this book, and a particularly important one as it takes on a number of tasks: it not only deals with a phenomenal orientation, as directed by my overall schema, but also with the parallel political development in popular modernity of, on the one hand, fascism, and, on the other, consumerism. In this way, the chapter helps to elucidate popular modernity's character as a managed or engineered cultural matrix. This chapter also deals with the images of popular modernity that were first developed during the establishment of corporate advertising culture in the 1920s, particularly those that are both spectacular and transgressive, from the horror stories of H. P. Lovecraft to the slasher films of the 1970s and 1980s. Two other transgressive image systems examined here are, first, hard-core pornography, which moved out of the red-light district and into the public sphere through the ascendancy of the home video player in the late 1970s and again through the Internet in the 1990s; and second, entertainments that have a strong reference to a fascist aesthetic, leading to a consideration of the ongoing significance of fascism in liberal democratic consumer culture. The chapter also looks at a broad range of related popular iconography: soft-drink advertising, body-builders, and rock music's "guitar hero" icons. All of these technologically mediated mythic images are examined in the context of the phenomenology of image perception, and, in particular, semiotic psychoanalysis of Self-Other and Self-boundary relations as they relate to iconography. Another important theme of this chapter is the role of transgression—or rather, imagined transgression—in popular modernity's overall management system.

The fifth chapter constitutes a case study that draws on two of our established phenomenological categories, voice and image, and applies them to late-twentieth-century developments in black American culture, whose unique status in American popular modernity rests in its dual character of being alterior, and yet very much at the center.

The problematics of Self and Other, racial/national identity versus class identity, the processing of cultural expression by the popular music and film industries, and, again, the phenomenology of voice and image are all examined with reference to rap music, which was first formed in a decidedly postindustrial New York City in the 1970s and which used sound reproduction technologies in decidedly new ways. For contrastive purposes, we'll look at heavy metal music, an allegedly white genre whose trajectory parallels that of rap.

The sixth chapter closes in on a matter that has been important throughout: narrative. To this end, I examine the story structure of popular modernity as found in sensationalized events such as the Sand Cave Tragedy (1925) (arguably the first modern media event), in the transformation of professional wrestling into a spectacular theatrical event, and finally, in the combination of mythic narrative, entertainment, and national politics that culminated in the Age of Reagan. I will also draw on one of Franz Kafka's parables, "A Hunger Artist," to provide a literary reflection of popular modernity as it operates in politics, news, and entertainment. But my endeavor here will be to examine not only some exemplary narratives of popular modernity but also to examine narrative itself in terms of its multiple status as text, mythohistory, phenomenal category, and a technology, and in so doing I will draw on the work of Martin Heidegger and Paul Ricoeur.

Because my interest is more with the human implications of the technologies and entertainments of the past 150 years than it is with any specific artifact, technology, genre, or medium, this study, as the outline above suggests, is more invested in breadth than depth; that is, I do not purport to present an exhaustive history of any given form of entertainment or media, or of popular modernity itself; rather, my aim throughout is to present the dynamics of the technology-myth-experience triad as revealed in a number of selected examples, thereby suggesting a paradigm for understanding popular modernity. In this effort, I have employed a wide rage of interpretive works, including, to name just a few of the more important sources, those that have traced the broad outlines of human cultural organization (Benedict Anderson, Guy Debord, and David Harvey), the realm of the intersubjective (Alfred Schutz), the interaction of the human subject and technology (Don Ihde, Carolyn Marvin), communications theory (Harold Innis, Marshall McLuhan), and the way in which specific cultural practices are historically and semiotically situated (Tricia Rose, Carol Clover).

The cultural and technological specimens—from Bing Crosby to Ice Cube, from *Brigadoon* (1954) to *House Party* (1990), from telegraph to Internet—have been chosen as apt examples (and of course there are many, many others) of the dynamics of technology, mythohistory, and experience in producing and reflecting popular modernity in America.

1

American Technospace and the Emergence of Popular Modernity

Imagining the Nation

Consider the kind of maps that were, until very recently, generally used in primary and secondary education in the United States: maps that tear the Eurasian land mass in two so that North America can have the privilege of being imaged at the center of the world; maps that render South America as about the same size as Greenland, whereas in fact it has nine times the land mass of that ice-covered isle that is both misnamed and misrepresented; maps that prioritize political boundaries over natural ones, thus rendering North America as a neat stack of brightly painted boxes (little irregular ones on the right, big rectangular ones on the left) rather than, to adopt another perspective, a system of geographic regions. The mythohistorical power of these maps is evidenced in one simple and absurd example: every year a fair number of tourists will drive well out of their way just to stand at the only place where four state borders (Arizona, Colorado, New Mexico, and Utah) meet. The "game" of standing at the spot where the "four corners" meet is a kind of real-space enactment of a popular elementary school activity as described in a 1937 geography primer, in which pupils create cardboard maps of the continental United States, cut it up along state boundaries, shuffle the pieces, and put it all back together—"the one who does it the fastest wins" (G. Miller 1937, 13). We are talking here about a particular way of thinking about space.

But what is space? Or more to the point, *where* is it? The mathematician Carl Gauss held that whereas number is a product of mind,

1

space has a reality outside the mind, and thus its laws defy a priori description (qtd. in Innis 1951, 92). This position, informed as it is by Euclid (space as a chartable domain, defined by axioms and geometric postulates) and Newton (space as a container; "absolute space . . . always similar and immutable" [Kern 1983, 132]), largely accounts for the accomplishments of positivist history and geography. It is a position that compels us with its "common sense"; even those who are well aware of the various and complex forms of subjective and inter-subjective spatiality ultimately succumb to "an overarching and objective meaning of space which . . . in the last instance . . . is pervasive" (Harvey 1989, 203). Nevertheless, we may want to quarrel with this view, as does Edward Said when he states that space is "something more than what appears to be merely positive knowledge" (1978, 55)—a statement that echoes the counterpositivism of Leibniz, Kant, Piaget, Husserl, Bachelard, and others who have posited the "whereness" of space, to one extent or another, within the human subject.

If we draw a distinction between the process of "mapping" and the product that results, "maps," we can conclude that the former is an imperative of consciousness—the need to situate ourselves within a world that is beyond our immediate perceptual reach—while the latter is a technology that radically expands our ability to do so by providing representations of the world in which we live. Maps originate in their use-value in terms of our desire to situate ourselves in the world and to operate effectively within it. But in the natural attitude in which we use maps to extend our vision and abilities, we forget that they are founded on a contradiction. As Denis Wood remarks, maps give us a "reality beyond our reach . . . a reality we achieve no other way. We are always mapping the invisible or the unattainable or the erasable, the future or the past, the whatever-is-not-here-present-to-our-senses-now and, through the gift that the map give us, transmitting it into everything it is not . . . *into the real*" (1992, 4–5).

Consider also the broad variety of uses for which we employ our maps. In some cases, as with a nautical expedition following the first set of exploratory maps of a coastline, or in the case of the lost and exhausted motorist examining the flashlight-illuminated road map while searching desperately for the interstate sign on the side of the rain-swept highway, the user will consult the map as he or she actually views the terrain, thus forming a kind of primary relationship of self-map-world. In other cases, as in geography education, maps have a very different and more secondary relation to the user, who will probably never visit the places he or she sees on the map and will know these places and peoples in an indirect and inaccurate way. In this latter usage, maps serve, not an immediate use-function, but a secondary and mythohistorical one: that of assigning a given hermeneutic value to the world (both spatially and temporally, though our focus here is on the former) beyond immediate apprehension, telling us what it *means*. In the former instance, the use-value of the map is more iconic in nature, of value for its immediate resemblance to that which it represents, while in the latter, the map is closer to text, to Saussure's "sign proper" in its arbitrary and conventional character and in its relationship to cultural signification in the form of the abstract mythic narratives that Barthes calls the "second order semiological system" ([1957] 1977, 114–15).

The post-Renaissance map, whose mode-of-presentation we very much take for granted, is characterized by its fixed, elevated, out-of-reach viewpoint (Edgerton, qtd. in Harvey 1989, 244). This is not at all a "natural" way of looking at the world (although it is a very "useful" one), for perhaps there is no "natural" way of looking; this map, like all mimetic practice, is culturally and historically coded and "loaded," as just one comparative look at a modern (i.e., post-Renaissance) map side by side with a medieval map, with its emphasis on the way terrain is experienced (as in Matthew Paris's Itinerary Map 1253, showing a narrow strip of land along a particular route), quickly demonstrates (cf. Harvey 1989, plate 3.3). The progeny of these Renaissance maps are the geopolitical maps used in compulsory education, and they add another layer of seemingly "natural" meaning—a more overtly political and ideological meaning—to the spatialization of the world and thus to the way the perception of space is acculturated.

The map developed by Flemish cartographer Gerhard Mercator in 1569 was designed specifically for navigation, and for this reason, it renders compass directions as straight lines. The Mercator, which makes the land masses north of the equator appear much larger than those to the south, was eventually transferred from one use-context, navigation, to another, the pedagogical one of providing young people with a view of the world; ultimately, the Mercator became the most widely proliferated, and therefore most immediately recognizable, world map. We can regard this as a mere accident of history: the Mercator was familiar and available, and it was innocently transferred from one context to another as its intended use was forgotten, a process that describes the evolution of all sign systems, like words whose original metaphoric value has worn away with use ("coins which have lost their pictures and now matter only as metal, no longer as coins," as Nietzsche says in one of his notebooks from the early 1870s [1954, 47]), or as with the maypole, whose original ritual symbolism became mystified and simultaneously mundanized as it became a mere toy. On the other hand, is it a "mere coincidence" that the world's most popular, recognizable, and familiar map "shows Britain and Europe . . . as relatively large with respect to most of the colonized nations?" (Turnbull and Watson 1993, 7). After all, maps with superior use-value in terms of conveying relative land mass were available: as early as 1570, the "sinusoidal projection" world map, which preserved relative land mass, was available; in 1772, Johann Heinrich Lambert proposed a "cylindrical equal-area projection"; and in 1855, James Gall devised an alternate world map. (It was almost identical to one designed by Arno Peters [who was apparently unaware of the Gall map] in the early 1970s, and is now known as the Gall-Peters Projection.) And in 1925, J. Paul Goode developed the "homolosine equal area projection" (commonly compared to a flattened orange peel), which has the added value of constantly reminding the viewer that he or she is looking a flat representation of a spherical object (cf. Monmonier 1995, 9–15). Better maps, then, were available; as Monmonier demonstrates, there was no shortage of criticism of the Mercator map, perhaps stated most strongly by U.S. State Department geographer S. Whittemore Boggs in a 1947 issue of *Scientific Monthly:* "[T]he use of the Mercator projection for world maps," Boggs declared, "should be abjured by authors and publishers for all purposes" (qtd. in Monmonier 1995, 21). While, as Monmonier reveals, the Mercator has largely disappeared in educational and responsible commercial outlets (as in the Hammond, Rand-McNally, and National Geographic publications), it continues in popular culture, "thriving" in the form of wall maps, promotions, and "cheap atlases and encyclopedias occasionally sold in supermarkets" (22). Certainly its mythic power, like that of the color-coded political boundary maps, continues.

What this demonstrates is simply that the pedagogical use of maps is ideologically interested. And this is true not only in the case of maps that distort the world to one's own ideological advantage: it is also true of attempts to make maps more "accurate." A good historical example of this is found in the first concerted initiative for compulsory geography education with an emphasis on "accuracy," which was taken up by the French after their humiliating loss to the Prussians in 1871. In the years following the Franco-Prussian War, the French government concluded that insufficient mastery of geography on the part of their field commanders played a key role in the defeat, and the responsibility for the rectification of this educational shortcoming was delegated to the Ministry of Public Instruction, which responded vigorously, designing and implementing a new geography curriculum (Graves 1975, 42–49). Other nations followed suit, and in the late nineteenth century there was a general flourishing of geography education. The effect of politics and ideology on geography education (and ultimately, on the ideology of space) is also well-illustrated in the American sphere. In the last decade of the nineteenth century, under the influence of Harvard professor William Morris Davis and with the mandate of the National Education Association, American education began a concerted move towards physical geography, which remained a strong element of compulsory education until World War II. However, the quality of geography education was inconsistent, a source of irritation for critical educators like G. Stanley Hall, who in 1911 declared that school geography was a sloppy mixture of disciplines, written by generalists who lacked true geographical knowledge and driven by textbook profits rather than scholarly merit. Hall declared that geography was the "sickest of all sick topics in the curriculum" and an expression of both the mediocrity of American pedagogy and "the character of our people, who crave to know something, but not too much, of everything . . ." (1911, 555–56).

After World War II, the social studies movement virtually destroyed whatever advances had been made in geography pedagogy (and, as Hall's commentary reveals, those advances were at best partial, limited, and inconsistent). While the advanced study of geography—regional or areal differentiation in the 1940s and 1950s, and later the rise of human geography—continued in the universities, in compulsory public education the discipline became largely absorbed by civics and history in an effort to promote patriotism and the idea of the United States as a model for the rest of the world. And thus it is that we come to the map of brightly painted boxes, disproportionate continents, and Winnebagos full of westering vacationers and retirees stopping for a photo opportunity at "The Four Corners" in the American

Southwest to thus photochemically inscribe their virtual injection into the American geopolitical map.

Another way in which space is ideologically constructed is found in the case of geographical directions—places "Other" than where we are. According to Said, "The East" emerges in the European imagination not as a positive geographic entity, but as an imaginative space that signifies, among other things, "insinuating danger . . . [where] rationality is undermined by Eastern excesses, those mysteriously attractive opposites to what seem to be normal values" (1978, 57). Such a concept of the East—and, by extension, attendant racial stereotypes—is abundant throughout Western intellectual history and is readily found in popular culture and, again, in pedagogical practice. In the earliest American geography "grammars," written by the father-and-son team of Jedidiah ("the father of American Geography") and Sidney Morse and in use from 1784 to at least 1828, racial and ethnic typecasting is considered part of the legitimate scope of geography. Thus, J. Morse's *Geography Made Easy* (1784) tells pupils that Spaniards are "lazy, proud, cunning, and revengeful" while Swedes are "grave, self-opinioned, and distrustful." (Relatedly, Morse used his *Geography* to "square away the untidy aspects of Puritan myth" [Seelye 1998, 151–52]). Morse's example was not ignored: shortly thereafter, in Nathaniel Dwight's *A Short but Comprehensive System of the Geography of the World* (first edition 1795, many editions thereafter), we learn that the Irish are "vehement," Turks "morose, treacherous, passionate, [and] unfriendly," and New Englanders "the most intelligent people in the world." The latter sentiment accorded with the elder Morse's heroic rendering of the New England settlers, a practice that has informed myths and rituals of the American settlement to the present day (qtd. in Brigham and Dodge 1933, 3–8; cf. Seelye 1998, 152). Similarly, B. Franklin Edmands's *Boston School Atlas* (1832) divides nations into four categories: savage, barbarous, civilized, and enlightened, with the final group exerting "the greatest and best influence on mankind," and of course "the United States and some parts of Europe are of this class." The *Boston School Atlas* speaks "the regularity and symmetry of their features" in describing the "Caucasian race," whereas other races seem to be a distortion of norms; they are described as having "thick lips," "flat noses," and "projecting foreheads" (Edmands 1832, 18).

The assignation of ideological value to geographical entities is particularly evident in the American notion of "the frontier," which serves less as a historical "reality" than as an index of a particular ideological orientation toward the world. If, as Spengler contended, each culture's conception of space is its "prime symbol" and an informant to its every aspect (qtd. in Kern 1983, 138), then perhaps the substratum or "prime symbol" of all American myth and ideology is the notion of "frontier." Frederick Jackson Turner's widely influential frontier thesis (*The Frontier in American History* [1893]) employed U.S. census data from 1890 to postulate that an ever-advancing American frontier, was, until its final closure in the late nineteenth century, the engine of American history. Turner's thesis reflects a "scientific" approach to history—an attempt to impose an empirical discourse on the terrain of political space as well as on the discipline of history (K. Klein, 1997, 14; cf. McNeill 1986, 3–22). The ideological underpinnings, however, of Turner's thesis are evidenced in both his postulation of a binary historical mythos that pits "civilization" against "savagery" (a received idea, one we saw in the 1832 Boston *Atlas*) and in his definition of a specifically American frontier, which unlike a European frontier (a border between populated regions) refers to a line between populated and unpopulated "free" land. The equating here of "savage" and "free land" reveals that Turner's historical explanation works only if we assume that the Hispanic and Native American peoples who occupied North America before the Anglo conquest and expansion did not do so in any legitimate way. Thus, Turner's thesis placed any questioning of American imperialism under erasure, suppressed by a quasi-empirical historical discourse in the service of myth, that is to say, a mythohistory par excellence (Turner 1893, 203). The mythos of the Turner thesis was apparent from the very beginnings of the Anglo-American project in 1620, when William Bradford, who would become the first governor of the Massachusetts Bay Colony, bemoaned the "wild and savage" (62) aspect of the new land and typologically invoked the image of Moses on Pisgah in his account of the arrival of the Pilgrims in *Of Plimouth Plantation*. More accurately, the "historical" landing at Plymouth initiates mythohistorical thought through the formation of what anthropologist Victor Turner calls an "ideological communitas" (qtd. in Seelye 1998, 9), while later, as Seelye painstakingly demonstrates throughout *Memory's Nation*, the event itself becomes the object of reverential and sentimental mythologization, as in Charles Lacy's famous engraving, *The Landing of the Pilgrim Fathers* (1850). We find further evidence of the development of the idea of

Manifest Destiny in the first colonial maps: in the 1612 map of the Virginia colony, as in John Speed's 1627 *A Prospect of the Most Famous Parts of the World* and Augustine Herrman's 1673 map of Virginia and Maryland, native peoples are represented as little more than decorative elements, literally in the margins of the map (King 1996, 105). Thus, the Eurocentric biblical interpretation that constituted the myth of Manifest Destiny became in Turner an empirical justification, which would later, in the realm of popular entertainment, be transformed back into mythohistoric narrative, particularly in the tales of the American West that have long been a staple of television and Hollywood film.

Turner's writing is a good example of the workings of the social and ideological construction of political space, an activity determined not so much by physical reality (e.g., empirical space) as it is by other ideological processes and constructs, both prior and ongoing. That is, the American notion of frontier is a "fact" of social history: for the American "frontier" suggests growth and opportunity, and furthermore informs a whole range of cultural postures, as in Kennedy's political program (the New Frontier), or the exploration of space (the Last Frontier), or the post–World War II move to the suburbs (the Crabgrass Frontier). But the grounding of this social condition is mythohistory—the redemptive myth of a continuing frontier, which is, Elazar notes, "the source of renewal that sustains the United States as a 'new society'" (1994, 75). Or as Shames states, the "fantasy of empty horizons and untapped resources has always evoked in the American heart both passion and wistfulness" (1989, 30–31). And, armed now with these observations regarding geography education, Turner's thesis, and the transformation of both Puritanical (biblical-mythic) and Turnerian (empiricist) discourse into that of popular modernity, we may revisit the "Four Corners." The tourist may observe this spot with reference to the American political map (the point at which Arizona, Colorado, New Mexico, and Utah meet) and feel a sublime sentimental nostalgia for the "open West," with its the erasure of the Other. (The "Four Corners" is also part of the land of the Navajo Nation, the largest and most populous North American Indian reservation, and by that Other conceptualization, there is no "corner" there at all.)

Eric Hobsbawm, in his discussion of the "invention" of the modern nation-state (France in particular), postulates that the three most important strategies in the invention of nations are public ceremonies, public monuments, and public education ([1983] 1994, 77). Maps are a technology that emerges from, first, a fundamental phenomenological impulse (the sense of emplacement in a surrounding world), and second, from an immediate use-value. In the context of "inventing" political entities and mass loyalties, their use-value is shifted from an immediate one to a mythological one, a process that is completed when the map becomes naturalized. An example of this is found in the notion that north is at the "top" of a map; such an idea is of course purely conventional, for in the largest geospatial context, north may be a magnetic pole, but it is certainly not the "top" (Turnbull and Watson 1993, 6–7]; nevertheless, this is a "natural" idea, and one can quickly demonstrate this "naturalness" by simply looking at a world map "upside down," which will inevitably strike one as being "wrong." Thinking back to the white face as it was described in the 1832 *Boston School Atlas* (which here stands as an example of a widely accepted, naturalized view) we see a parallel: the white face is normal and "symmetrical," while black, brown, and yellow faces are distortions; therefore, the white face is like the map "right side up," with any other form of representation constituting an abnormality.

The nation-state, from its origins, has a particular relationship with cartography, as it does with other technologies—with, for instance, literary production and communications technology. According to Benedict Anderson, the political order of the modern world could only happen when, for a variety of reasons, the great "transcontinental sodalities" (Christendom and Islam) were no longer ideologically viable ([1991] 1994, 89). It is here that technoeconomic change plays an important role; improved printing technology in the sixteenth and seventeenth centuries created a new permanency to the national vernaculars, and the era of "print capitalism" set the stage for the rise of the modern nation-state. The "rise of the novel," as Watt suggested, mirrored the trajectory of capitalism in the early eighteenth century, but it also served to help the nation-state to congeal; the novel (and the newspaper) helped to "standardize language, encourage literacy . . . [and] remove mutual incomprehensibility," and generally abetted the intersubjective, psycholinguistic encoding of national identity—in short, the nation is imaginatively conjured with the aid of novelistic discourse and the technoeconomics that support the culture of the novel (Brennan 1993, 48–49; cf. Anderson [1991] 1994, 35). There is, thus, a relationship of necessity between communications technology and nationalism. Karl Deutsch grasped this singular insight best, and he made it the sine qua non of his definition of nation: the national group (a "people") is "a larger group of persons linked by . . . complementary habits and faculties of communication" (1953, 96).

However, the ways in which technologies "naturalize" the nation depends on the specific character of the technologies themselves. Seton-Watson identifies the older nations as those that had "acquired national identity or national consciousness before the formulation of nationalism" and the newer nations as those in which nationalism among the masses was engineered by the self-conscious direction of a revolutionary leadership ([1977] 1994, 136). But in the United States and Canada, the historical foundations of the "old" nations do not fully apply, and the conditions of national formation are somewhat different than in the "new" nations, for in North America (what we might call the "new new" nations), industrial technology to some extent precedes culture, particularly if one regards North American national formation as taking place in the mid-nineteenth century rather than in the closing decades of the eighteenth century. In North America, newer technologies were employed in the process of imagining nationhood. The railroad and the telegraph reencoded ideological space by assisting the ongoing formation of a unitary political identity—a belief in the United States as a unified spatial field and, hence, unified ideological field.

The emergence of the telegraph in the 1840s played a special role in the technoeconomic reencoding of geopolitical space, and it seems particularly fitting that the telegraph's inventor, Samuel Morse, was the grandson of Jedidiah Morse, whose role in geography education we have already noted. It is reasonable to assume that the younger Morse, through constant exposure to the concepts of distance and space that preoccupied both his father and his grandfather, developed his interest in the space specifically as it relates to communication (that is, space is inevitably a communications barrier) (Blondheim 1994, 30). And indeed, the ultimate importance of telegraphy—historically, socially, and phenomenologically—is rooted in that primary relationship between space and communication and in the way in which telegraphy radically modified this relationship through the phenomenon of time-space compression.

There were a number of pre-electronic methods for sending high-speed messages. In the ancient world, the use of fire signals to organize military campaigns amounted to a kind of "ancient telegraph" (Hersh-bell 1978, 81), and as Crowley notes, the ancient Greeks used polished metal and reflected sunlight to send such messages; in Africa there was the talking drum, in North America, the smoke signal; in early-nine-teenth-century France, the mechanical semaphore. The telegraph how-ever, because of its scale, marked a leap from a *transportation* model of communications to a *transmission* model; for the first time, transporta-tion and communication were truly separate. This quantum techno-logical transformation meant that for the first time not only could information move independently of and faster than physical entities, but it could control the future movement of commodities (Crowley and Heyer 1991, 124). This in turn, Carey contends, informed a parallel economic shift from arbitrage (speculation based on spatially separate regional markets) to futures (in which space collapses and speculation is based instead on *time*—the possible future value of a commodity) (1991, 135).

Whereas Americans were becoming accustomed to the conquest of space and time entailed by the steamship and the locomotive, the disjunctive shift from the transportation to the transmission model of communication presented new conceptual difficulties. As the telegraph was not used by any large number of people directly or in a domestic-use context (as would prove to be the case with later technologies), it was not, in any direct way, part of the everyday life-world. We can, however, witness the reactions within the smaller social circles of the power elite to this new technology. The very uncanniness of the tele-graph, as Blondheim reveals, was such that its invention was probably more easily accomplished than its acceptance by the legislators and business leaders whose support was needed in order for Morse to ob-tain the needed start-up capital. In the earliest demonstrations of the telegraph, the power brokers Morse wished to court suspected that they were the victims of an elaborate hoax. The difficulty was in con-vincing them that "the clicking machines they were watching were actually responding to operations taking place miles away," and the general suspicion aroused by the device is well-illustrated by the fact that the first appropriation bill for the telegraph presented to the U.S. Congress in 1843 was encumbered by a rider for funds to support mesmerism research (Blondheim 1994, 31–32).

As Blondheim notes, in 1844, Morse set up a demonstration de-signed to win over the skeptics: he successfully telegraphed the results of the Whigs' national convention in Baltimore to Washington, D.C., twenty-two miles away. When the conventioneers arrived in Washington nearly two hours later, they confirmed what had already been telegraphed—that the relatively unknown Frelinghuysen has been chosen as Henry Clay's running mate, a development that no one, much less Morse and his assistant, could have known (Blondheim 1994, 31–32). Significantly, this public demonstration of the validity of the device established that the telegraph would be subordinated to the older print technology: it would be a handmaiden technology that would only be used by a new class of information technicians, the telegraph operators, in the service of the newspapers and news agencies. The effect, then, of the telegraph, in terms of mass intersubjective experience, is secondary in nature. The older form of nationalizing media, Anderson's "print capitalism," mediated the new media, thus providing an instance of what I will call hypermediation (the process by which one medium directs the reception of another).

While the novel, Brennan observes, accompanied the formation of the nation-state by "objectifying the 'one yet many' of national life" (1993, 49), the newspaper was just as important (perhaps more so) in the United States, with its limited literary output in the early nineteenth century; this is particularly true when we turn to journalistic representations of the telegraph during its first decade. The journalism of this period abounds with laudatory manifestos devoted to the telegraph. One of the most florid practitioners of the telegraph disciples was James Gordon Bennett, editor of the *New York Herald*, the best-selling newspaper in the United States at that time. According to one of Bennett's columns from 1844, the telegraph would "blend into one homogeneous mass . . . the whole population of the Republic. . . . [It could] do more to guard against disunion. . . . than all the most experienced, the most sagacious, and the most patriotic government, could accomplish" (qtd. in Hietala 1985, 197). The metaphor that soon emerged imaged the railroad and telegraph as the muscles and nerves of the national body (an ironic conceit, given that the railroad lines, at the local level, would soon become the standard boundary between white and black communities, thus dividing the nation). Bennett was also a strong supporter of the war against Mexico, and indeed, he went even further than most pro-war spokesmen by advocating the American conquest of *all* Mexico, not just the northern provinces of California and Nuevo Mexico. Bennett's two political positions (pro-telegraph, pro-war) were not unrelated: the imagining of nationhood entails not only metaphors of national unity, like the railroad and telegraph system, but also the imagining of the Other who stands beyond the national boundary and who is often perceived as a threat to it. Indeed, the connection I have been exploring here between the conceptualization of space, technology, and political ideology is borne out by the general tenor of Democratic Party rhetoric during this expansionist period, for the remarks of the party leadership often betrayed their feeling that the "conquest of distance was as important . . . as the conquest of Indians and Mexicans" (Hietala 1985, 197).

The older print media, we see, metaphorically rendered the telegraph in terms of national unity, and in this example we come to face with the methodological problem of developing phenomenal descriptions of technological experience, a problem rooted in the complexity of social and textual constructivism vis-à-vis technophenomenology. Providing phenomenal descriptions (or trying to get at such descriptions through historical accounts) is problematized to the degree that the response to a given technology has already, to some extent, been directed by another medium (usually a previous and already familiar one), which in turn provides a mythic framework for the conceptualization of new communications technologies.

The next major development in communications technology was telephony. Its establishment and history demonstrate a continuity with telegraphy, but it also contributes to the development of popular modernity in ways that go well beyond its primary construction (e.g., populist politics and media guidance of popular opinion) during the Age of Manifest Destiny.

The telephone was initially conceptualized in terms of its only existing analogue and predecessor (the telegraph, of course), a conceptualization that was probably abetted by the fact that many of the early organizers of telephonics had begun their careers in the telegraph industry (this accounts for the fact that the telephone's first imagined use was that of allowing telegraph operators the ability to talk to one another [Lubar 1993, 119]). Again, we see that new and initially uncanny technologies come to us only through the mediation of older, naturalized technologies. More significantly, however, telephony demonstrates a continuing pattern of new technologies being mythopoeticized and ideologically co-opted through hypermediation. The older print medium rendered telegraphy as part of a set of mythic signifiers related to imperialism, national superiority, and Manifest Destiny, and the telephone (and later, as we shall see, broadcast radio) was likewise drawn into such a process.

First of all, telephony served as a national unifying device; by 1915, AT&T public relations advertising used a tactic that harkened back to the jingoistic journalism that had accompanied the establishment of the telegraph some seventy-five years earlier by using a map of the United States in the ad and phrases like "the telephone unites the nation"; the telephone was "the welder of the nation" that made "the continent a community" (Fischer 1992, 163). But here we find a difference. After the establishment of AT&T in 1885, the telephone was promoted largely as a business tool, and only later as a tool for the facilitation of household business (shopping, making appointments, attending to emergencies, etc.). But by the late 1920s—owing largely to the use-patterns (*consumer* use-patterns) that had developed largely beyond the control of AT&T and to the capital lure presented by the prospect of expansion into the general residential market—the telephone became increasingly conceptualized and marketed as a social facilitator (Fischer 1992, 41, 79). In this way, telephony marks a decided turn away from a producer orientation and toward consumerism. Returning to our example, then: the difference between the imagining of the United States in relation to telegraphy and what we find during the establishment of telephony is that the latter participates in what we might call the *commercial* imagining of the nation, a merger of the politics of consumerism and the older politics of nation-statism that, in a new user-based technological environment, in part defines popular modernity. In a series of advertisements for Cremo cigars, for instance, the image of the continental United States is conjured by clouds of cigar smoke, a weird conjoining of national identity and consumerist oral gratification, and thus an example of popular modernity par excellence (fig. 1). As for the relationship between technology and national consciousness, there is some evidence that the telephone may have initially strengthened local ties socially; but many experts (Kern, Westrum) see the telephone as "yet another of modernity's blows against local *Gemeinschaft*" (Fischer 1992, 23).

Perhaps more accurately, telephony reencoded this *Gemeinschaft* through a "decentralization of an urban lifespace into a matrix of intimate social networks" or "psychological neighborhoods" (Wurtzel and

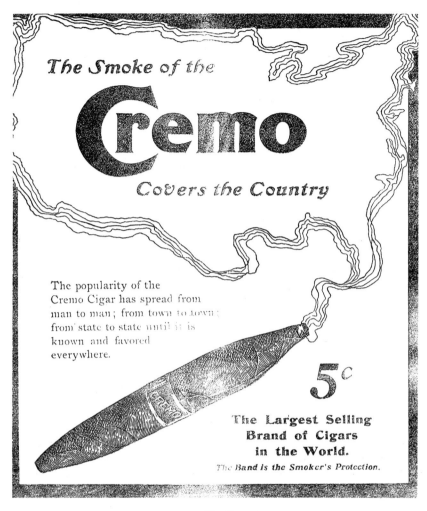

Fig. 1

The Commercial Imaging of the Nation. Advertisement for Cremo Cigars (1903),
Smithsonian Institution American History Archives, Ayer Collection.

Turner 1977, 246; cf. Aronson 1971 and Ball 1968). While there is considerable range of interpretation here, it seems clear that one way or another, telephony created a new spatiophenomenal nexus. In other words, while the social and phenomenal impact of the telegraph was largely restricted to that which could be achieved through its secondary hypermediation, telephony (and the technologies that would follow) would have both a hypermediated *and* a nonmediated relationship with the everyday life-world and on concepts of self and the emplacement of self in an ambient spatial field. Because of its far greater integration with the everyday life-world and the habitat, telephony reencodes the phenomenology of spatialization in a way that initiates the "continuous sensory and spatial reorganization of social life" (Berland 1992, 43) that is perhaps popular modernity's most salient feature.

In relation to the individual subject, telephony (unlike its predecessor) is implicated in what Ihde, drawing on Heidegger, calls embodiment relations, "a symbiosis of artifact and user within a human action" (1990, 73). The telephone becomes an extension of self—in using it repeatedly, the human subject becomes increasingly unaware of its presence, thus making it not so much an intermediary between self and world, but a part of self in what is experienced, in the natural attitude, as unmediated experience. (Ihde uses the example of eyeglasses, which at first are experienced as an alien presence and later seem to become part of one's self.) In like fashion, the telephone becomes an extension of self/voice.

But this new kind of technological self-extension, a kind of cyborgization, while potentially empowering, can also have the opposite effect based on an opposing phenomenological condition, for it entails a situation in which others may encroach upon one's boundaries. These boundaries, based on both sensible and conceivable elements of the real environment, are defined phenomenally by a series of concentric (and egocentric) circles that provide a sense of enclosure and emplacement, and, as a derivative, protection and belonging. We can gather from Husserl that the midpoint and the outermost circle of this concentric system are, correspondingly, the human body (corporeality) and a much larger body, the Earth, "perceived in a primordial synthesis as a unity of mutually connected single experiences . . . yet, it is a body!") (1981, 222). But for our purposes in this immediate discussion, we are better occupied with the intermediate circles affected by telephony—that is, the house or the immediate dwelling, followed by the community or city (Jager 1985, 215). Regarding the first of these two: our rooms become our wombs: in our personal habitats, Bachelard suggests, "memory and imagination remain associated" in a way that is indexed to our childhood and a feeling of "motionless security" (1969, 4–5).

The telephone is essentially an appendage to the habitat that is a potential threat to the womblike security or some other habitat (or fortress, if we think of the patriarchal image of the man's home as "castle"), because it gives the walls of the habitat a kind of porosity. As Fischer notes, a common complaint during the first two decades of telephony was that it "permitted intrusion into the domestic circle by solicitors, purveyors of inferior music, eavesdropping operators, and even wire-transmitted germs" (1992, 26). If we think of the two primary locations of the telephone in the home, the kitchen and the bedroom, we can see that this technology has been metaphorized in terms of the two sets of social relations it extends and modifies: the domestic (with its associations with the hearth) and the intimate. To demonstrate the social and cultural concerns that were generated in terms of the latter, we need only refer to the kind of prurient interest in the possibilities of "phone sex" that began within the first decade of commercial telephone with the "hello girls" who served as personal alarm clocks for male subscribers. Some of the commentary from the popular press of the late 1880s evidences a kind of connection between the telephone and sexualized relationships, capitalizing on the popular myth of a triangle between husband, wife, and telephone operator (Marvin 1998, 106–7). Strangely enough, telephony was already sexualized and hypermediated before it was electrified: du Moncel, in an early work on telephony, makes reference to a "string telephone" that, "if we may believe some travelers . . . has long been used in Spain for the correspondence of lovers" (1880, 12), and his book contains a wood print of two young courtiers using this device in an apparently amorous manner. Thus the relationship between communications technology and sexual behavior has a long tradition, from the lover's telephone to the various forms of anonymous sexual discourse made possible by the Internet (cf. Turkle 1995). Alexander Graham Bell, in the deposition he gave pursuant to the suit brought forward against him to annul his patents, mentioned that he had bought a device in Boston that had long been known as a "lovers' telegraph" (1908, 211).

Also, some of the earliest advertising played on this theme, such as a humorous set of cards created by a Canadian Bell company in which a protective matron cuts the telephone wire with a scissors to prevent two young lovers from using the phone for sweet talk: "There! That'll stop their nonsense!" she gloats triumphantly (Fischer 1992, 166). The telephone, to return to our "hello girls," thus seemed to provide transgressors with the ability to penetrate the physical perimeter of the domicile for the purposes of stealing the sexual property within.

The presence of these largely subliminal and intersubjective (psychological and phenomenological) concerns were energized and brought into full public consciousness much later in the Supreme Court case of *Olmstead v. U.S.* (1928), the first case to really probe the issue of domestic privacy vis-à-vis electronic technology. In this case, the Court, opting for a narrow interpretation, determined that the Constitution had not been violated when federal agents tapped Olmstead's telephone wire, since they did so without physically trespassing on his property (clearly, eighteenth-century notions of space were inadequate, and as is usually the case, the law lags behind technology). In a dissenting opinion, Justice Brandeis warned of the awesome power of technological espionage and concluded that the intent of the Constitution was to "protect Americans in their beliefs, their thoughts, their emotions, and their sensations. They conferred, as against the government, the right to be let alone" (qtd. in Long 1967, 23). If privacy is to some extent a modern invention related to the private experiences afforded by literacy and reflected in private reading habits, letter writing, and the keeping of diaries, then the breakdown of this privacy is a development stemming from the inception of electronically mediated popular modernity.

Not surprisingly, then, in the early controversy regarding the telephone we find a site of tension in the relations between the family and the notion of domestic privacy, the nation-state, the corporation, and technology. As we review its evolution, it also becomes evident that the modern (consumerist, atomized) family evolved from the family ideal formulated during the establishment of the bourgeois nation-state, a model that was disseminated across class lines and through a series of social changes. As Rupert Emerson noted, the nation is "the largest community that . . . commands . . . loyalty, overriding the claims both of the lesser communities within it and those that cut across it or potentially enfold it within a still greater society" (1962, 95). According to this view, the family unit is an entity that the nation-state relies upon as a cell that reproduces national values and ensures that the need for workers, consumers, and military conscripts will be met. As industrialization and the centralization of labor eliminated the middle-class woman's role in commodity production, it was restructured along the lines, suggested by Coventry Patmore's poetic ode to marriage, of the "angel in the house" (cf. Auerbach 1992, 66); as Mosse puts it, the Victorian domestic ideology proposed that the home should be a "warm nest into which one could retreat from the pressures of the outside world," into a world of privacy, comfort, and the absence of conflict (1985, 18–19). And these post-laissez-faire interventions into the domestic sphere continued and continue to support this structure, for as Barbara Nelson noted, American family assistance, starting with the New Deal, was built on two tracks of assistance: male breadwinner and female wife/mother (qtd. in Coontz 1992, 138).

The cultural neurosis regarding the telephone as an intruder is the result of the way in which it can abrade this domestic ideology. In other words, the ideology of domesticity that was established by the bourgeoisie as a response to their own industrial revolution and furthermore encouraged, as we noted, through the privacy expectations generated by broad literacy, was now being interfered with, at least symbolically, by one of the technological products of that revolution. This perception regarding the telephone continued for quite some time; indeed, as late as the 1960s, when the size and power of AT&T was just beginning to come under attack, there was considerable public concern regarding the telephone as an invader of privacy, as reflected in a number of films, such as *If A Man Answers* (1962) and *I Saw What You Did* (1965) (Lubar 1993, 139–40). This sense of the telephone as an invader of the home (a charge that would later be directed at radio, television, and the Internet) is perhaps why ultimately the telephone became a "fossilized" technology; that is, in spite of the fact that the picture-phone has been possible since the 1960s, the public seems to have become resistant to any major changes in the telephone, and modems, cellular and "smart" (i.e., computer assisted) phones, and consumer services (like three-way calling) aside, very little has really changed regarding its use. Perhaps this is because there is no interest in extending the invasion of the habitat to the visual level.

The Screen/Space

In a number of fanciful illustrations that appeared in the popular press
in the early 1880s and in which all the eventual uses of television,
including distance education, entertainment, and even home shopping, were imaginatively presented (Barnouw 1975, 4–7), it is evident
that the dream of television existed in the popular imagination long
before such a device became technologically and commercially viable.
At about the same time, Paul Nipkow began to design a "visual radio," and its perforated, rotating "Nipkow disk" would serve as the
technological basis of television experiments for decades, until it was
decisively replaced by the cathode ray tube. Of all modern technologies, television has had the longest technology lag, largely because it
did not suggest an immediate political and military application (as did
radio) and because its development and mass deployment was waylaid
by two world wars and an intermediary period of economic depression.

With the saturation of American domestic space with television
sets, a process that began in 1947 and was largely completed by 1955,
a radically new form of spatiality became part of mass culture and
consciousness. First of all, as with the radio before it (which will be
discussed at length in the next chapter), television as a physical object,
a household appliance in "real" domestic space, posed certain problems rooted in the basic contradiction between what Ihde calls the
"framed space," which is, through the very act of framing, out of the
ordinary (1996, 126), and, on the other hand, the mundanity of the
domicile space. As Spigel points out (1990), when the television was
first introduced into the domicile space there was an attempt to cam-
ouflage it with "hideaway" cabinets. Television came to occupy a
designated space within the American domicile only when it was "incorporated within an overall furnishing scheme" and assumed a role as
an ancestral shrine, as a place for "treasured objects, such as family
photos and mementos" (Morley 1995, 182). But the spatial meaning
of television, with its total "flow" (R. Williams 1974, 86–96) of moving images, always available and, unlike cinema, fully "domesticated"
in the home setting (cf. Barthes 1980), goes considerably further than
its object-status in domicile space: we need to consider the nexus of
real and virtual spaces in reference to television's generic programming.

The genre that seeks to directly represent the space of the American home—the situation comedy—provides a good example. As Ozersky observes, the typical spatial presentation of the home-space in the situation comedy is such that the television is missing. When the fictional television family watches television, they sit facing us: "[W]ith the screen nowhere in sight, the illusion exists for a moment that the TV really is . . . a mirror or reflection of us" (1991, 11). A mirror, we will recall, is not only a functional device but an architectural one as well, often used to make a living space appear larger, and in some sense, the architectural function of the television has been to do just that—to phenomenally expand our living space. This leads us to consider another popular television format, the talk show, which over the past twenty-five years has gone from "talking heads" to an audience-participation mode established by Phil Donahue and then developed in various ways by Oprah Winfrey, Geraldo Rivera, and finally, the carnivalesque Jerry Springer (cf. Howe 1999). This format effectively capitalizes on the illusion of three-dimensionality, for whereas the older talk shows presented two flat figures engaged in conversation, the Donahue format relies on the host energetically moving from the stage platform to the studio audience area. The studio audience, not unlike the chorus of Attic tragedy, represents "us," the television audience. "They" are intended to represent some sort of central moral voice as they confront the "guests" on stage, who increasingly are intended to represent deviance from normal behavior. As Tomasulo notes, "[T]he at-home audience . . . must delegate its own talk power to the show's participants, especially the host or hostess [who] 'stand[s] in for' the absent audience and ostensibly brings clarity to the complicated issues of the day." In so doing a form of "alleged democracy is put into place that seemingly represents . . . the struggle between competing currents of opinion within the culture" (1994, 6). The studio audience members are "us" transported into the tube, and they ask the participants the hard questions that assumedly "we" ourselves would ask—in short, they *mirror* us. And in a sense, the entire format effectively expands private space, for the mirror trick of television allows us to incorporate the studio—which is decorated in living-room fashion, with plants and sofas and so forth—into our own, more meager, habitats.

If we turn our attention to a third form of television programming—the Western—we will be able to probe aspects of the American ideology of space beyond (yet enclosed by) the domestic sphere. Westerns were presented in television series beginning in the early 1950s. There were popular series like *Gunsmoke* (1955–75) and *Have Gun Will Travel* (1957–63), and shows featuring mythohistoric figures like Annie Oakley, Jim Bowie, Wyatt Earp, and Kit Carson. From their inception, Westerns enjoyed steady growth in prominence and popularity, and by 1959 they accounted for slightly more than 24 percent of prime-time programming. Some series drew as much as 32.5 percent of audience share, and they provided symbiotic support for Hollywood Westerns. No other form of programming, Slotkin points out, "commanded so consistently high a share of prime time over so many years" (1992, 348). In the process, it promoted the taste for the nostalgic in popular entertainment, aimed here specifically at a mythic frontier past. This nostalgic taste would later be considerably "foreshortened," as will be discussed in depth in the third chapter—but returning to our spatial emphasis here, this nostalgic programming had a relationship to the phenomenology of time-space compression. A stereotypical image that characterized both the television Western and its mediation in television shows and movies depicting television and televiewers is that of the U.S. cavalry going into battle against the Indians on a broad and open plain—which seems peculiarly penned in by a nineteen-inch rectangular surface, as if the programming were an attempt to dispel the phenomenon of spatial compression both signified and phenomenalized by the television itself. Ironically, television—which, in Meyrowitz's view, eliminates emplacement by bringing other spaces into the home (1985, 124–25),—was used in this instance to stimulate the nostalgic longing for an imagined national space ("the Wild West"), a nostalgia that began with Turner's thesis, with the postulation of a "moment" quantified by the 1890 census at which, in purely positivistic terms as reckoned by population density, the frontier had closed. In the cultural imagination, this purportedly scientific calculation takes on mythic proportions. Further, the function of the television Western is similar to that of the maps of North America and its frontier first produced in the seventeenth century; the television Western is a kind of electronic, ideological map. One of the most popular and long-lived of these programs, *Bonanza* (1959–73), which rode high on the ratings charts in the early 1960s, featured an opening sequence in which a map of "the Ponderosa" (and the Spanish name of the mythical home of the Cartwright family is one of the few, though largely invisible, references to the American West's Spanish heritage) burns away, like Baudrillard's (1994) Borges-inspired "map that covers the territory," to reveal the images of the hero-family galloping on horseback through their ranch lands.

Our understanding of the three popular television genres noted above as well as the experience engendered by television programming in general can be considerably expanded if we draw on Foucault's notion of the heterotopia. The best metaphor for the heterotopia is provided by that complex and strange yet simple and familiar object, the mirror: a "virtual space" underneath the surface; a place that puts me "over there, where I am not," in a kind of "shadow that gives my own visibility to myself." It is a place of both ultimate reality and ultimate unreality: "[I]t makes this place that I occupy at the moment when I look at myself in the glass at once absolutely real, connected with all the space that surrounds it, and absolutely unreal, since in order to be perceived it has to pass through this virtual point which is over there." And finally, perhaps most significantly, it is a place where "all the other real sites that can be found within the culture, are simultaneously represented, contested, and inverted. Places of this kind are outside of all places, even though it may be possible to indicate their location in reality" (Foucault 1986, 24). Quite obviously, our experience of television is heterotopic, whether we are talking about situation comedies that "mirror" our living rooms, talk shows that phenomenally expand the domestic space to bring in "guests" who mirror, in a broader way, received ideas about normality and deviance, or the typical Western program circa 1957–63, which served the purposes of the dominant national ideology by compressing the mythic "open spaces" of the Old West and refracting them into domestic space.

It is strange that although television programming, video games, and the Internet share a strong experiential homology (i.e., they are all activities centered on the viewing of a framed screen/space), the available commentary and analysis, both academic and mass market, generally fails to explore the continuities of this linkage. The linkage, however, is easily demonstrated in the rectangular shape and aspect ratio (width:height ratio) of the computer screen. The aspect ratio of the word-processor screen that now glares before me is almost exactly the same as that of my television at home, while the aspect ratio of the conventional printed page is almost exactly the opposite. Thus, the word-processing computer's spatial presentation has more to due with the confines of cathode-ray tube production and televisual entertainment rather than with the print media's conventions. The linkage of television and home-computer use did, however, eventually come back together in 1998 as retail outlets like Circuit City promoted the personal computer as a way to watch television through the use of enabling software.

In spite of this, the available commentary has been more focused, not on the continuities, but rather on the discontinuity in the management of screen/space in the shift from television broadcasting to personal computing. The video game, developed at MIT in the early 1960s and first marketed by American firms in the early 1970s, reached its full cultural significance in the early 1980s through the success of the Japanese firm Nintendo, which brought the era of totalizing North American hegemony in media culture to an end. Video gaming brings new qualities to television use: if television's "total flow" is an outpouring of material to be passively consumed (or sucked down like mother's milk in the womb-security of the domicile, as Houston suggests), then the outfitting of the video screen with interactive software and control panel transforms that passivity into interactivity. According to this view, the video game remasculinizes the "feminine" passivity of television, particularly for the adolescent males who were the prime target of the video game movement during its heyday in the mid-1980s. Kinder elucidates the masculinist and oedipal nature of these programs (often manipulated with a "joystick") by suggesting that their popularity rests on the fact that they were introduced to the U.S. market "at a time when fewer households included a father," and thus these "games provided an appealing surrogate against which a son can test his powers" (1991, 103). Fiske agrees, observing that the popularity of these games in urban video arcades among subordinated males is based on their ability to provide a space in which to "think through, and rehearse in practice, the experiential gap between the masculine ideology of power and performance and the social experience of powerlessness," thereby providing a fictive space in which the subordinate males receive "awards and recognition that they never do in society" (139). And thus, we witness the odd phenomenon of one phenomenal style masquerading as another—a technology that *presents itself* as a Heideggerian, tool-like self-extension while actually being the technoheterotopia par excellence, a place for the playing out of psychodynamic fantasy. Thus, given the rather limited parameters of the game-con-struct, perhaps the interactivity of the video game is really, to enact an Adornoesque coinage, a kind of pseudointeractivity.

For the reasons above, it is tempting to suggest that such games do not gear into the world in any real way—but then there is the example of the Persian Gulf War, in which, it has been suggested, the success of American soldiers operating "smart weapons" can be traced to their previous practice on video games, thus turning the oedipal fantasy of the game toward a geopolitical oedipal struggle with, as Levidow sug-gests (1995, 9–15), Iraqi president Saddam Hussein as the evil Other/father, which in turn became a totalizing form of television entertain-ment. This was, Kellner observes, the first war that was "orchestrated for television" (1992, 110), with dramatic images of soldiers and tanks with the Arabian Desert as backdrop and new, Windows-style com-puterized presentation techniques (cf. Caldwell 1995) and an attrac-tive program title ("Desert Storm"). Even General Schwartzkopf was compelled to remind his troops on one occasion that war "is not a Nintendo game" (qtd. in Lubar 1993, 276). Further, there were re-ports that some of the tank soldiers experienced trauma when, on emerging from their tanks to examine the results of their campaign, they realized that the war was conducted in real, geopolitical space, not in hyperspace. In his discussion of this phenomena, Kevin Robins argues that in what was a "push-button, remote-control, screen gaz-ing war" the video screen was the "only contact point, the only chan-nel for moral engagement with the enemy other." For the viewer at home, this "amplified and legitimated the sense of omnipotence and power over the enemy" (1996, 79). The difference between the soldier's use of the video screen and that of the citizen at home watching an account of the war on CNN is the difference between interactive video and traditional television use. Video, used in this way, is part of a self-extending embodiment relation, while television functions as what Ihde calls a hermeneutic relating technology—a technology that tells us how to interpret what we see through its manner of presenta-tion. It is, as Ihde implies, television's seemingly natural "body-like" seeing, its "seeing is believing" power, and its guided use as a herme-neutic relating technology that makes it particularly prone to political uses. In the Gulf War, to further elaborate the point made by Robins above, the televisual presentation of the technological capabilities of the U.S. armed forces, a purportedly neutral, fact-based presentation, conceals the hermeneutic code that encourages an implied conclusion: one of American moral superiority, technical superiority, and righteous omnipotence over a racial Other. This is not very different from the cultural climate of the 1840s, the age of the telegraph and the war against Mexico.

Whatever one might claim about the differences between "traditional" televisual experience (including the postbroadcasting or "narrowcast" cable programming, which, Marc claims, ushered in a new period of television in the early 1980s [1984, 167]) and the televisual activity of video gaming, Internet surfing, e-mailing, playing in the computer network interactive "multi-user" domains (MUDs, MOOs, and MUSEs), we need to consider these two historically situated forms of televiewing in terms of the way intersubjective expectations of these technologies are created through the process of hypermediation and the way they are forestructured by a nationally formed ideology of space. In order to do that, we need to do some backtracking.

The telegraph was described by the media and politicians alike in masculinist terms: with the tandem-technology of the railroad, it helped form the muscles and nerves of the national body. Some of that masculinist impulse can be seen in an account in the *Reading (Pennsylvania) Herald* from 1889. It provides a fanciful description of the new New York to Boston long-distance telephone service. A "Hello!," it says, passed through the telephone, out of the room, out of the building, into a conduit, and then,

> rising along under the turmoil and rush of New York City, then up the Hudson, taking a squint at the Palisades, past Yonkers and Tarrytown and Sleepy Hollow, . . . [went] through New Haven, Hartford, Providence, Newport, on to Boston. It crossed rivers and mountains, traversed the course of fertile valley and past busy factories . . . before it plumbed into [Boston] and reached its destination in the ear drum of a man seated in a high building there. . . . (Qtd. in Marvin 1988, 196)

The rhetoric of this passage conveys a vigorous sense of physical space and its conquest by technology, and thus, in this specimen from the daily liturgy of journalism, we see a transference from the rhetoric of telegraphy to that of telephony. It is furthermore interesting to point out that the piece was written in 1889, just one year before the census report that claimed that the physical frontier had closed, and just four years before Turner's famous essay, and it is therefore tempting to suggest that in this account we see the transference of the energy of the frontier mythos into a new and more technological domain, yet still rendered in terms of a physical geography. The journalist's fanciful description of the "hello" moving through the wire reads like a paraphrase of "God Bless America," in which the human subject seems to whisk over the American landscape (viewing mountains, prairies, and so forth). The telegraphic muscularity of this particular hypermediation of telephonic communication would be later reencoded, in the evolution of the use-practices and the marketing of the telephone, into a different kind of contact—the domestic and sentimental contact summarized by the slogan, "Reach out and touch someone."

What, then, of the hypermediation and the intersubjective conceptualization of computer networking? It is common practice now, a practice that I too shall follow, to link the evolution of the conception of the postbroadcasting screen/space to science fiction, and particularly to novelist William Gibson, who coined the term "cyberspace" after observing the obsessive involvement of video game players, an involvement with a virtual space behind the surface of the screen (interestingly enough, the first video game, developed at MIT in 1962, was called "Space Wars"). In *Count Zero* (1987), Gibson prophetically described cyberspace in terms of a "consensual hallucination, " a place of "unthinkable complexity" where cores of information burn "like . . . novas, data so dense you [would suffer] sensory overload if you tried to apprehend more than the merest outline" (38–39). Gibson's imaginative rendering owes much to Burkeian and Kantian ideas of the sublime (Voller 1993, 21–22)—ideas that in turn informed American notions of the frontier as presented in that period's landscape painting. An interesting reinvention of such spatial visions is found in popular graphic renderings of cyberspace, such as the images that accompany *Time* magazine's 1996 special issue, "Welcome to Cyberspace," which is illustrated with colorful artistic interpretations of blue, hairless, sexless bodies in free-fall flight through a cyberworld of institutional knowledge (representations of schools, museums, government buildings, etc.). It is noteworthy that at the time of the publication of that particular issue, and as reveled therein, a Yankelovich poll revealed that 57 percent of the American public were not familiar with the term "cyberspace." Thus, these illustrations, appearing as they do in one of the most popular news magazines, clearly forestructure and hypermediate the conceptualization of this technology. And there are many other, similar popular images and rhetorical usages. Between 1994 and 1996, over three hundred articles were published in the United States that used terms like "electronic frontier," "cyberspace," or, in keeping with the theme of taming the savage wilderness, "cybercops." In the related realm of product advertising, there was a vast quantity of promotional material from manufacturers like Northern Telecom, Honeywell, IBM, Apple, and Intel, first, in the early 1980s, for "information age" business equipment, and, by the early 1990s, for the domestic market for "home" computers (cf. Salvaggio 1987, 151).

It is also important to note how these hypermediations change while retaining a degree of ideological consistency. Descriptions of Internet communication generally lack the vigorous sense of motion through physical space that we find in our *Reading Herald* specimen describing long-distance telephony (for physical space has been rendered "irrelevant"), but the ideological orientation that corresponds to the signifier "frontier" has not disappeared; and thus, cyberspace, rather than geographical space, or outer space (which has lost much of its charm in the public imagination through its naturalization) becomes the repository for this peculiarly American sign. Or in Robins's formulation, these images and rhetorical formulations constitute a continuation of the utopianism that "has played a fundamental part in the modern social imaginary" and the attendant desire to suspend "the principles that regulate human existence in the mundane and real world" (1996, 15, 16).

And of course, in addition to this sublime romanticism and utopian idealism, there are also forms of nationalistic romanticism, which was abundant in the journalistic accounts of telegraphy (fanned as it was by the jingoism that was generated by the war with Mexico) and which have resurfaced in the hypermediation of computer networking; we see therefore that the same ideological forestructuring that accompanied the advances in communication technology in the 1840s are alive and well a century and a half later. This is evidenced by the conservative group affiliated with former Republican senate majority leader Newt Gingrich called "Democracy in Virtual America." The group referred to cyberspace as the "land of knowledge" and said "the exploration of that land can be a civilization's truest, highest calling" (Elmer-DeWitt 1995, 6).

Bob Metcalfe, a commentator for *Info World* and a developer of Internet technology, muses that "maybe we shouldn't be calling it cyberspace" (1994, 63). Another term that raises questions is "surfing": marketing campaigns have capitalized on the freedom and adventure implied by this term, as in a 1995 IBM commercial in which a nun proclaims that she "can't wait to surf the Net" (implying that religious rapture will be replaced by technological rapture?). Here the contiguity of television and computer networking is evident in the parallel between "channel surfing" and "Net surfing," the former accepted as an "activity" associated with boredom and listlessness while the latter is supposed a thrilling exploratory activity conducted in a free imaginative space—although *cyber* meant in Greek "to steer" or "control"— an etymology worth noting (cf. Metcalfe 1994, 63). In any case, in his discussion of "cyberspace" Metcalfe raises an interesting point: Internet activity is essentially the act of sending a message through a medium to a person on the receiving end (cyberspace may be thought of as being "exactly where you are when you're on the phone with your mom" [63]). This is certainly the case with e-mail, and when clicking onto a listing on a web browser, we are essentially ordering information at a given location to be sent to us. And yet, these activities are generally thought of as an act of "going" to some other place, a heterotopia. As Metcalfe suggests, this could be just as true of a telephone conversation, since this idea of going somewhere is merely figurative, as is the term cyberspace itself, all of which illustrates the influence of received ideas on the conceptualization and the phenomenalization of a medium. Stone likewise finds the term "cyberspace" problematic, for its "immense conceptual/gravitational field" as a definition has dominated "all the elements of virtuality . . . including those that do not particularly relate to it," elements that have, in any case, been with us for a long time in the form of "dioramas and botanical gardens, aurally as radio dramas, kinesthetically as carnival rides, textually perhaps as novels" (35).

In the ascendancy and universal acceptance of the term "cyberspace" we see further evidence of the suppressed relationship between conventional television and home computer activity. Because of the American public's thirty-odd years of television culture, there is a strong predisposition to think of the screen as a "window" to other places;

hence the success of the Microsoft Windows, introduced in 1985 as a way of giving computers with MS-DOS the "feel" of the Macintosh interface style, which, unlike the IBM style, emphasized the phenomenology of screen/space (Turkle 1995, 37). Internet activity is also logically and experientially linked to television (and the entire history of electronic communications) because it continues the phenomenon of time/space compression that began with the introduction of the telegraph and was continued by broadcast radio and the introduction of the television into the domestic sphere. By carrying messages digitally rather than analogically and relying on geosynchronous satellite transmission, time-space compression has taken a quantum leap. As Sola Pool notes, "[T]he communications distance between all points within [a satellite's] beam has become essentially equal. In the non-Euclidian communications plane that results, no point lies between two other points" (1990, 65).

Just as it is ironic that the tight, penned-in screen of the television set is used to convey narratives about the mythic, sublime, open spaces of the American West, it is ironic that computer networks implode space and thus make the world "smaller," and yet, this activity is metaphorized in terms of sublime spatial boundlessness. It is also important to note, however, that the computer is being spatially domesticated, just as the television was. By 1994, theoretical architects had moved into the design of virtual spaces (cf. Anders 1994), a move later reflected in the advent of software management systems that imitate the space of the domicile. An example would be Microsoft's 1995 program, "Bob," a "social interface" that presents interrelated utility programs (calendar, mail, checkbook, address book, etc.) in a graphic space that "looks like the living room from a '50s sitcom" (Strauss 1995, 68). The user views or grasps (clicks onto) items on the simulated desk or shelves, or the wall clock, to enter various the various programs. Strauss's guess—that Microsoft's motive was to create a program that would reduce computer phobia by domesticating the image of the home computer—is an accurate one.

In any case, the presence of the notion of boundless cyberspace side by side with "domesticated" programs indicates the two poles of American experience that forestructure the commercial introduction of this technology: the frontier mythos and its binary counterpart, the domestic mythos. If there is anything new in this latest wave of technology, it would be, as Robins suggests (1996), the rejection of physical emplacement in favor of the utopian space of virtuality. This undermining of physical location, which has been an ongoing part of the development of communications technology and which reached a new intensity, as Meyrowitz argues (1985), with television, reaches fruition in the ideals of cyberspace. In an age of postphotographic images (i.e., digital images that do not have the immediate connection with quotidian reality that photographs do), cyberspace seems to have left the "real" world behind. In the 1990s, there have been relatively few signs of critical resistance to the idealization of cyberspace and virtual reality; on the contrary, the media, the corporate world, the government, and even that medieval institution and a bastion of emplacement, the university, have issued manifestos to herald this glorious future. For example, John Lombardi, president of the University of Florida, looks forward to the day when the "cyberversity" can finally "let go of our current practices and truly float free into the brave new cyberworld" (qtd. in Pratt 1994, 50). In our quest to understand this current reencoding of space, we would do well to follow Robins's advice by interrogating the "expression of dissatisfaction with the real world . . . and with others" (1996, 17) that is implied in the ideology of cyberspace.

We have, to sum up, canvassed a sequence of technologies—maps, telegraph, telephone, television, the Internet. Phenomenologically, these technologies interact with the subject in three fundamental ways: they extend, become embodied with, and contain the self. But if they help us to expand our reach, our vision, and emplace ourselves within the larger world, they also, in their role as hermeneutic technologies, tell us how we should use them and delimit how we should see and how we should think. These technologies are thus a contested domain between subjective experience and the power elite's use of established mythohistoric discourse to forestructure that experience.

2

The Disembodied Voice:
Coughlin, Crosby, and Other Crooners

Having examined to some extent the phenomenology of spatial experience as it operates in the technological and ideological matrix of popular modernity, we now turn to a second basic phenomenal category: that of listening and voice. Here we shall again have to work toward descriptions that negotiate the analytical *aporia* that results inevitably from the experience:culture problematic I discussed briefly in the introduction, and in so doing, we will again identify instances in which hypermediation plays a role in naturalizing technological experience. We will also probe the role of mythohistoric construction in the critical work of Innis and McLuhan, in much the same manner as in our discussion of Turner's frontier thesis and the laudatory journalism that accompanied the establishment of the telegraph system.

The phenomenal problematic of vocality is quite different from that of spatiality. Whereas in the latter case, as noted earlier, there is a compelling commonsense predisposition to think of the space of the world as external and noncultural, which leads to a kind of blindness regarding the "interestedness" of, for instance, geopolitical maps, the human voice is, in the popular imagination, given a romantic rather than an empirical credit. Perhaps this stems from a combination of three things: first, the atavistic equating of "breath" and "soul"; second, the "aural fingerprint" quality of individual voices; and third, the voice's simultaneous functioning as a conveyor of both language and emotive qualities not necessarily linked to linguistic signs. It is also illustrative of the association that links the voice with music that, while no one would enjoy trying to read printed text in a language he or she does not understand, it is not at all uncommon for people to enjoy vocal music in unfamiliar languages, as in the case of a monolingual English speaker who enjoys the sound of Italian-language opera, or, to use a contemporary popular culture example, the sound of Spanish-language *tejano* music. This kind of popular practice is mirrored in philosophic analysis: in Husserl, the voice is given a special status derived from his notion of the "twofold sense" of the linguistic sign—*Ausdruck* (indication) and *Anzeichen* (expression). To apply this to our musical example, one who enjoys vocal music in a language he or she does not understand defaults the aesthetic pleasure of listening from indication to that of expression; Dunn and Jones are onto the same thing when they suggest terms like "embodied voices" and "vocality" in order to "indicate a broader spectrum of utterance," for often the voice manifests itself through singing, crying, and laughing, "each of which is invested with social meanings not wholly determined by linguistic content" (1994a, 1). Derrida's critique of Husserl centers on the question of voice: Derrida argues that Husserl's position in this regard is a repetition of the privileged notion of presence in Western metaphysics, and he rejects transcendental phenomenology's attempt to "protect the spoken word" (1973, 15). But even if Derrida's critique of Husserl is correct, it is only partially so, for Dunn and Jones's probing of "vocality" shows that the phenomenon indeed exists, but that it might better be probed through a postphenomenology ("invested with social meanings") rather than through classical (i.e., "transcendental") phenomenology. In any case, my point here is not to settle the issue of Derrida contra Husserl, but simply to reveal the controversial status of voice, which is better probed, particularly for my purposes here, in realms other than that of language philosophy.

Voice, Technology, and Power

A controversial thesis regarding the power of voice is found in Julian Jaynes's *The Origin of Consciousness in the Breakdown of the Bicameral Mind*, in which it is argued that prior to the dawning of consciousness as we know it, there was an epoch of "bicameral civilizations"—a time when individuals did not reflectively think out and "narratize" problems, but rather responded obediently to instructions that may have been delivered by members of the priesthood through the hidden speaking tubes that were part of many stone idols, such as the Head of Orpheus, a major ancient oracle at Lesbos. These instructions, once internalized, controlled behavior through auditory hallucinations. (Given this, and to jump ahead for a moment, it is interesting to consider the corporate icon for RCA that first appeared on 78 r.p.m. phonodiscs: a dog peers down the barrel of a phonographic amplification trumpet, its head inquisitively cocked. The caption: "His Master's Voice.") Jaynes supports this thesis in part through an analysis of the representation of subjectivity (or rather, the lack of subjectivity) through its reflection in the form of heroic age narrative. In Jaynes's reading of *The Iliad*, its characters

> do not sit down and think out what to do. They have no conscious minds such as we say we have, and certainly no introspections. . . . It is the gods who start quarrels among men that really cause the war, and then play its strategy. It is one god who makes Achilles promise not to go into battle, another who urges him to go, and another who then clothes him in a golden fire reaching up to heaven and screams through his throat across the bloodied trench at the Trojans, rousing in them an ungovernable panic.

Ultimately, Jaynes concludes, the gods "take the place of consciousness" (1976, 72).

In addition to this literary deduction, Jaynes draws on empirical findings from the field of brain physiology: while all our control of speech (in the right-handed subject) is in the *left* hemisphere of the brain (in what is known as "Wernicke's Area"), there is nonetheless an area in the *right* hemisphere that is capable of organizing speech and that can assume this function in cases in which the tissue in Wernicke's Area is damaged. Jaynes postulates that this parallel area, which is connected to Wernicke's Area through the anterior commissure that connects the two hemispheres of the brain, "may have organized admonitory experience and coded it into 'voices' which were then 'heard' over the anterior commissure by the left or dominant hemisphere" (104).

Jaynes's argument that auditory experience was of greater magnitude in preliterate societies gains some support from the etymology of the verb "to hear." In Greek, in Latin, in Hebrew, in French, in German, in Russian, and in English, "hear" and "obey" share the same root (from the Latin *ob* + *audire* = to hear facing someone [Jaynes 1976, 97]). Conversely, *eidomai* in classical Greek means both "to see" and "to know"; and in our vernacular, "I see" means "I understand." These etymologies (as well as the weight of received wisdom, as in Aristotle's claim that "sight is the principle source of knowledge" [Ihde 1976, 82]) take on particular significance in the epoch of popular modernity.

At the beginning of the modern scientific era, developments in the technology of lens crafting and the science of optics allowed us to observe new worlds, from the planets of our solar system to microscopic life forms. As Ihde points out, however, this "extension of vision," while quite obviously transforming and enlarging both our empirical knowledge and, in turn, our conceptual frames, has also "*reduced* [our] experience. . . . for the world which began to unfold through the new instrumentation [of the telescope and the microscope] was essentially a *silent* world" (1976, 5–6). We, however, most certainly do not live in a silent world. The shift from the silent world of early rationalist science to the world of electronically reproduced sounds and voices—the shift from optics to electronics, as Ihde puts it—has meant that for the past century or more, since the late 1870s with commercial telephony and since the introduction of the microphone, loudspeakers, and then broadcast radio in the 1920s, we have been a culture that spends a good deal of its time, like the schizophrenic, listening to disembodied voices. Further, if, as Jaynes argues, listening and voice were phenomenal categories of privileged cogency in preliterate societies, we need to consider that this may be true too of postliterate society.

In the opening decades of telephony, it was customary to present the new invention at public demonstrations; at one of these events held in California in 1897, a naive subject summarized the experience thus: "Although I had seen a talkaphone before I had never tried one before this morning. I could hear the voice very clearly although the speaker was a long distance from me. It was like a voice from another world. Here I was speaking to a person far away from me whom I could hear as though he was at my side and yet I could not see him" (qtd. in Fischer 1992, 62). Those who first heard these etherized voices via the very first broadcast transmissions—at General Electric on Christmas Eve, 1906—reacted similarly; in the words of one: "a human voice coming from their instruments—someone speaking! Then a woman's voice rose in song. It was uncanny!" (qtd. in Barnouw 1975, 20). In other cases, cases we should regard as extreme, some reported feelings similar to seasickness resulting from listening to voices through the telephone (Kittler 1990, 23), while in other cases, as in one from 1890, the "aural overpressure" caused by the telephone was judged a cause of insanity (Marvin 1988, 132). Yet another type of commentary that warrants interest is provided by a journalist for the *New York Times* reporting on the role of wireless telegraphy in saving 745 human lives in the wake of the 1912 sinking of the *Titanic:*

> [F]ew . . . realize that all through the roar of the big city there are constantly speeding messages between people separated by vast distances, and that over housetops and even through the walls of buildings and in the very air one breathes are words written by electricity. (Qtd. in Kern 1991, 187)

These accounts tell us much about primary experiences with these voice-disembodying technologies as well as the ways in which journalistic accounts hypermediate their conceptualization and thus contribute to a culturally/ideologically coded attitudinal stance toward the technologies themselves that must be factored in as a modifier of primary experience. If we take Ihde's concept of "embodiment relations"—the incorporation of an initially uncanny technology, like one's first pair of eyeglasses, into the self—we can conclude that new communications and entertainment technologies quickly become naturalized. A parallel example is found in the initial experience of motion pictures, which, according to Hugo Munsterberg in his 1915 study, were capable of causing "sensory hallucinations" in "neurasthenic persons" (qtd. in Gabler 1998, 50). In any case, both the telephone and the movies quickly became part of mundane experience, what we might call the "naturalized uncanny." We see, furthermore, that telephony's embodiment relation has additional complex intersubjective and diachronic dimensions. Consider the example of a child born after the dawning of the telephonic age and who has yet to master the codes of speech, but who still is able to exchange a few words over the telephone with his grandmother and seems to take the whole process quite naturally. It is no more strange than speech itself, and we see that primary experience in such cases is culturally and socially forestructured.

But returning to the previous examples: in the accounts given by ordinary subjects (ranging from awe to nausea), we detect the uncanniness produced by a first encounter with telephony, while in the journalistic specimen, the rhetoric suggests a shift from the uncanny to the grandiose, for there is clearly an attempt to impress upon the reading public the importance of the radio medium and its power over space and time. The shift, then, is from an expression of a subjective and disinterested experience to one that is imbued with interestedness and intersubjectivity in the form of a public mythology. In this way, the mythologizing journalistic ethos is a kind of intermediate step between primary experience of the technological as uncanny and the fully naturalized experience, in which both of the prior experiences are pushed into the background, though there can be variations in the ordering. In the example of the child at the telephone, the subject may be struck by the uncanniness of the technology *after* the naturalized experience, and he or she is sure to come upon mythohistorical codifications later, at that juncture in public education when American children learn to associate technologies with cultural superiority. Marked with the signet of sensationalist discourse, journalistic accounts such as these, like the accounts of the telegraph in the *New York Herald* some eighty years earlier, are prose poems—odes to the Hertzian sublime and to the alleged superiority of that nation whose people have technologically conquered the invisible mysteries of nature. As we shall soon see, this early journalistic praise for the wonders of radio would later transform according to an ideological dialectic that views technology, in alternating fashion, first as savior, later as destroyer.

We turn now from telephony to broadcast radio. During the period of its establishment (roughly from 1923 to 1933) in the skein of popular modernity, broadcast radio developed rhetorical modes suited to its peculiarities as a new medium. Radio's rhetorical codes, Crisell observes, paradoxically derive, not from its advantages, but from its *disadvantages* as a medium with "no image and no text" (1986, 5). Examples of the use of these disadvantages are provided by Cantril and Allport in the first sustained study of the psychology of radio, in which they note that specific forms of "radio rhetoric" create an intimate experience (as in the forms of address used by radio's political and religious zealots; e.g., "my friend"), while at the same time a contrary emotive pull is provided by the "impression of universality" (created by phrases like "we are now broadcasting from coast to coast") ([1935]

1986, 21). This combination of intimacy and universality, contrary yet complementary and powerful in terms of rhetorical/psychological effect, was one source of radio's rapid acquisition of social, political, and cultural significance, and in its first two decades it accelerated the destruction of American regionalism, promoted a national dialect, and pushed further the uniformity of both national markets and consumer desires. But while the rhetoric of radio during this period of declamatory broadcasting was important, so too was the voice itself, which can— through accent, stress, pacing, modulation, and other expressive codes that, in the natural attitude, we regard as noncoded and therefore evidence of "presence"—form "so powerful an expression of personality that [it] can impose a unifying and congenial presence on the most miscellaneous of magazine or record programmes" (Crisell 1986, 43). The limitations of the medium—its textlessness and imagelessness— allowed its most skilled practitioners to draw more fully on these pretechnological characteristics of the voice as well as its modification through the use of the microphone, and create a metaphysics of presence as an operative ideology in popular culture.

In the United States, protofascists like Louisiana governor Huey P. Long and Father Charles Coughlin had great success with the radio. Long's Share Our Wealth Society gathered five million supporters through radio programming, and Coughlin's success was even more dramatic: commercial broadcast radio was in its infancy when he began his *Golden Hour* program in 1926, and he had virtually no public recognition prior to his radio appearances; and yet, in a few months time, the "radio priest" secured a membership of eight million in his National Union, "the closest thing to the Nazi party . . . ever in the U.S." (Cantril and Allport [1935] 1986, 9). Coughlin's audience, Warren notes, were "more than a passive and atomized mass. They were forming a cohesive electronic community" (1996, 27). Coughlin's closest associates acknowledge that he was aware of the special qualities of the medium and that he had a particular ability to use the social psychology of identification as well as his unique Irish tonality (no doubt taken as a sign of old-world "authenticity") to advantage. His advisor, Louis Ward, remarked that Coughlin was "master of those hundreds of keys and stops which, when touched, played the melody of hope or a requiem of sorrow, upon that great organ, the human heart," and Coughlin himself said that radio must be "human, intensely human."

In his subtle understanding of the uses of technology, Coughlin pre-
figures both the politicians and the entertainers in the latter development of popular modernity in that he was one of the first figures to
break the barrier separating "public opinion formation and celebrity"
(Warren 1996, 24–27). Relatedly, the case of Coughlin demonstrates
one of popular modernity's most salient features—the conflation of
entertainment and politics, a subject we shall return to in the final
chapter in a discussion of the Reagan presidency.

The Coughlin phenomena repeats itself in the rapid rise to media
stardom of Rush Limbaugh (twenty-one million listeners in 1996),
Howard Stern, and other talk show radio hosts (and this may remind
one of Marx's dicta on historical repetition: first time as tragedy, second time as farce). Since the advent of television, broadcast radio had
come to be seen as a fossilized media technology, a "poor cousin of
the media" and "a sleepy, antiquated sideshow" (R. Turner 1996,
58). Yet, while radio has been succeeded by successive technological
waves—television broadcasting, cable, the Internet—it has actually
remained strong and has grown stronger with the ongoing expansion
of automobile commuting. This gives us pause to consider the transformation of the radio audience over the decades—from isolated, head-
set-wearing, crystal set listeners, to the domestic parlor audience, and
finally, to the largely isolated automobile commuter (on this note, the
ongoing resistance to carpooling may indicate that Americans have
naturalized the experience of the solitary commute—that they like to
be alone in the car's enclosed space).

While Jaynes approaches the problem of voice and authority from
a psychohistorical perspective, Harold Innis, while sharing Jaynes's
penchant for the vast historical argument, approaches cultural change
with a theory in which communications technology replaces the
Marxian economic base as superstructural determinant. Innis, whose
remarks coincide with Anderson's, suggests that during the era of the
Treaty of Versailles, the impact of widespread printing made itself
known in the weakening and elimination of imperial autocracies (e.g.,
the Austrian Empire) and the concomitant ascent of text-based, con-
stitutional, print-oriented political concepts, with its catch phrases
"democracy" and "freedom of the press." Innis argues, however, that
in the new age of communications "based on the *ear* rather than the
eye" a significant reversal took place that led to the undoing of the
stability established by the Treaty of Versailles and political self-deter-

mination. "The rise of Hitler," he argues, "was facilitated by the use of the loud speaker and the radio. . . . The spoken language provided a new basis for the exploitation of nationalism and a far more effective device for appealing to larger numbers" (1951, 81). History, as rendered here, seems to have come full circle, not from Edenic tribalism to utopian socialism, but from ancient autocracy (centralized, authoritarian, inflexible, and perhaps, bicameral) to early-twentieth-century fascism—from the stone idol issuing commands from a hidden tube to the Nazi project of establishing a new political regime with the aid of a new communications medium. McLuhan adds to this the notion that communications media have either a "hot" or "cool" character, and he claims that if television (cool) preceded radio (hot), "there would have been no Hitler at all" (1964, 261).

These reflections from the Canadian school are valuable in that they account for the historical development of communications in terms of its structural homology with the history of political ideology (from print democracy to radio fascism) as well as providing a phenomenal rendering (i.e., "psychic echo chamber") of the media/subjectivity nexus. It should also be noted that the criticism of McLuhan's technological determinism, particularly Theodore Roszak's pointed attack, needs to be put in the context of the way McLuhan sets up his argument: rather than a purely deterministic system, he sees both the preexisting, post–World War I German paranoia regarding *Lebensraum* (living space) and the presence in Italy of an illiterate, preindustrial population as being essential to the way radio helped evoke fascist sentiment from preexisting social conditions (cf. Holbert 1998).

Having said that, however, the Innis-McLuhan thesis is not beyond its own mythohistoric constraints. Consider first that when Innis remarks on the slogans associated with print capitalism, particularly the concept of journalistic freedom, we see that the press played a role in mediating itself, for the invocation of "freedom of the press" as a sacred mantra of capitalist democracy is, as it is generally received, a political principle, but it is also a kind of metajournalism—that is, journalism produces discourse that conditions the public reception and evaluation of journalism, and the institution defends itself by offering up its own "charter," to use Malinowski's term. In like fashion, the Innis-McLuhan position is both an evaluation of the significance of radio *and* yet another "interested" mediation: that of the literary and academic establishments. McLuhan had his ideological formation in the literary-critical establishment, and this needs to be considered when we evaluate his postulation that the "Other" media, a media generally seen as hostile to literary habits, is to blame for the decline of civilization—"the reversal of the entire direction and meaning of literate Western civilization" (1964, 262). It is for similar reasons that McLuhan fundamentally misapprehends fascism as being merely an atavistic return to ancient political orders; on the contrary, fascism's ideology of vitalism and the creation of a "new man" are inherently modern, a point that might be missed by liberal-democratic critics who "Otherize" fascism by characterizing it as being merely primitive (cf. Payne 1995, 10).

In toto, the ideological and phenomenal construction of the radio voice draws on the pretechnological apprehension of voice, the qualities of the medium itself, and the phenomenal feedback effects of its own hypermediation in a hall of mirrors of statements, evaluations, historical analyses, panegyrics, and condemnations. All of them come with deep-structured ideological presuppositions, such as the kind of binary imperative that leads to the construction of the other, like Turner's civilization:savagery binary. On that point it's interesting to note that Innis established himself as an economic historian by rejecting Turner's thesis, although we see that both he and McLuhan maintain its binary code by "Othering" a medium (radio) and a political ideology (fascism). A structuralist equation of the code enacted here would be as follows:

evil : primitivism : radio : fascism :: good : modernity : print : democracy

This, then, serves as a reminder that cultural critique is always implicated in the mythologies, even as it tries to demystify them.

Another evaluative position that needs to be considered is hinted at in the psychologism of McLuhan's notion of a "psychic echo chamber," the inner/aural experience of "voices in the head." This approach can be more effectively elaborated through recourse to the psychoanalytic theory of the superego vis-à-vis the radio voice. In Freud, the superego is essentially the disembodied voice of the father, and thus when a radio voice is that of a national patriarch, the disembodied voice as superego becomes implicated in a powerful synergism of political ideology, the subconscious, rhetoric, and technology. The

American patriarch during the ascendancy of radio was Franklin D. Roosevelt, whose political programs and personal career were at first supported, and later decisively rejected, by Coughlin. Like Coughlin, perhaps even taking him as a model, Roosevelt developed a powerful mastery of the radio, just as other American presidents achieved power through their understanding of media—Lincoln, of public oratory; Kennedy, of television; Reagan, of the whole range of popular culture and its mythic associations; Clinton and Gore, of pop music (as in their saxophone and guitar shenanigans during the 1990 campaign). Roosevelt, who spent "countless hours" developing his radio technique in consultation with staff writers and literary figures, most notably Archibald MacLeish (R. Brown 1998, 16), demonstrated his media mastery most successfully in his radio address of 5 March 1933 in response to the national banking crisis. The effects of the address were almost immediate; the White House was flooded with telegraphs that unanimously revealed confidence in Roosevelt and his policies (Cantril and Allport [1935] 1986, 210)— confidence in the disembodied voice of the father. If Coughlin and Long were closer to the authoritarian superego of Freud, perhaps FDR's mastery is better explained by Abram Kardiner's theory, in which the superego is not "the product of menacing repression caused by parental dictates," but rather "a category that society creates over time [that] constitutes the sensible, moral substratum of a democratic society" (Cantor 1988, 166). Much the same could be said of Phil Donahue's twenty-five-year dominance in the television talk show; for, as Tomasulo argues, he served as both the liberal and enlightened "new male" and father figure to his predominantly female audience, whose authority lent him "control of the means of production of Talk" (1994, 6–7). We should also be aware of the continuity in the radio/television transition, since "viewing" habits often indicate that television is also, and is perhaps primarily, an aural medium, since viewers routinely wander away from the set and listen, rather than watch, the television. [Morley 1995, 170). On the other hand, while we need to be attuned to the differences in political systems, we should avoid the knee-jerk reaction of trying to build a cordon sanitaire between fascism and welfare-state liberalism. On this point, it's noteworthy that Mussolini, in response to one of FDR's more fiery speeches, his inaugural address (which we'll look at further from the standpoint of narrative in chapter 6), praised the American president by saying that his inaugural speech demonstrates that

the whole world feels the need of executive authority capable of
acting with full powers of cutting short the purposeless chatter of
legislative assemblies. This method of government may well be defined as Fascist. (Qtd. in Ryan 1988, 83)

In a statement that valorizes command and authority as well as the
commanding voice, American radio commentator Edwin C. Hill said
much the same:

That voice, with its supreme confidence and courage, still resounds
in my ears. . . . Roosevelt. . . seized the helm of the leaky, drafty
ship of state with a hand that revealed no slightest trace of indecision
or of uncertainty. . . . Something had happened to the cross section
of the American people within the hour, . . . [f]or they had looked
upon, and they heard the voice of[,] a man who was willing to lead.
(Qtd. in R. Brown 1998, 62)

Having looked at the way radio broadcasting was valorized in
journalistic discourse (and later we will look at the way this journalistic stance would change after radio's first decade) and critiqued in
academic discourse, we must now examine the way radio was mediated by literary production. The problem of voice is central not only
to the technoeconomically driven popular modernity that we have
been discussing: it is likewise central to the modernist literary movement (Pecora 1985, 993). In Joseph Conrad's character Marlowe in
Heart of Darkness (1902) we find a man who was "all voice," and in
this voice lay his power and authenticity. Joyce, Stein, and, Faulkner
were also preoccupied with the speech patterns and the human voice
as rendered in literature. The case of Ezra Pound is of particular interest in this regard. In titling his magnum opus the *Cantos* (1926), Pound
announced that voice was central to what would seem a text-based
modernity. Further, Pound's support for Coughlin, Long, and Mussolini, and his own addresses on Mussolini's Rome Radio (which lead to
his arrest and imprisonment for "radio treason") reveal that this architect of literary modernism was drawn to the authoritarian voice of the
populist demagogue. "The figure of Mussolini," Redman suggests,
was "highly cathected in Pound's psyche" (1991, 98). T. S. Eliot, who
wrote specifically on the nature of voice in his "The Three Voices of
Poetry," also made the fascist connection in *After Strange Gods* (1933),

in which he argued for cultural and religious homogeneity, a United Christian Europe, and warned that "any large number of free-thinking Jews" was "undesirable" (qtd. in Ackroyd 1984, 201). Perhaps the modernist's preoccupation in this regard was related to the new way the human voice was being heard—that is, just as the emergence of photography encouraged painters to rethink the problem of light, the "Other" voice of radio encouraged a rethinking of literary voice.

For a specific modernist literary treatment of the radio and the voice, we would do well to turn to John Dos Passos, whose own literary significance may be described as the intersection of the modernist aesthetic (formal, stylistic) experimentation and the social realism that superseded modernism. Dos Passos employed self-conscious literary hypermediation in the form of his "newsreel" technique in his *U.S.A.* trilogy, and he repeated this technique with reference to the radio medium in an prose poem entitled "The Radio Voice" (published in *Common Sense* the year after the banking crisis). The piece begins with a rendering of the security of domestic space: we pan in on a small frame house during a furious rainstorm. The house is a "tiny neat stable wooden box full of dryness, warmth, and light," but it is full of "a lonely tangle of needs, worries, desires." In the parlor the family listens "drowsily to disconnected voices, stale scraps of last year's jazz, unfinished litanies advertising unnamed products that dribble senselessly from the radio." Suddenly, the programming is interrupted for an address from the White House. The listeners become attentive as they picture FDR, a man at his desk, leaning toward the listener, "speaking clearly and cordially to youandme," letting us know that he sits in the White House with his "fingers on all the switchboards of the federal government, operating the intricate machinery of the departments," writing codes, regulations, bills, all for the good of "youandme." This "youandme" sitting "worried about things . . . close to the radio" is defined by Dos Passos in an catalog that includes people from all walks of life: miners, farmers, mechanics, homeowners, farmers, consumers, bankers, mortgage holders, and so on, a catalog that concludes with the phrase, "not a sparrow falleth but . . ." FDR goes on to assure the listeners that recovery, including good wages, secure jobs, and protected bank deposits, is just around the corner. The listeners "edge . . . closer to the radio . . . flattered and pleased, we feel we are right in the White House," and when the voice stops

we want to say: Thank you, Frank; we want to ask about the grand-
children and the dog that had to be sent away for biting a foreign
diplomat. . . . *You have been listening to the President of the United States
in the Blue Room.* . . . No wonder they go to bed happy.

But what about it when they wake up and find the wagecut,
the bank foreclosing just the same, prices going up on groceries at
the chain stores, and the coal dealers bill . . . and that it's still raining?
(1934b, 17)

It is, first of all, significant that Dos Passos's impressionistic account—an account in which literary rhetoricity references radio
rhetoricity—of the Fireside Chat essentially parallels the combination
of overdetermined forces at play as deduced by Cantril and Allport,
for it accounts for the intimacy effect/affect ("sitting at his desk . . .
leaning towards youandme"), while at the same time accounting for
broadcast radio's declamatory "impression of universality" ("with his
fingers on all the switchboards of the federal government, operating
the intricate machinery of the departments"). And indeed, FDR's rhetoric was crafted to create this intimacy; in his very first Fireside Chat,
entitled "An Intimate Talk with the People of the United States on
Banking," every one of the five sentences of the first paragraph works
relentlessly toward establishing this relationship:

> —I want to talk . . . with the people of the United States. . .with the
> overwhelming majority . . .
> —I want to tell you . . .
> —I recognize that the many proclamations. . . should be explained
> for . . . the average citizen . . .
> —I owe this. . . because of the fortitude. . . with which everybody
> has accepted . . .
> —I know that when you . . .
> —I shall continue to have your cooperation . . .
> —I have had your sympathy and help . . . ([1933] 1938a, 61)

Dos Passos tells us, then, of the image that we're supposed to see when
listening to this I-thou rhetoric: the image of FDR at his desk leaning
toward "youandme."

Secondly, Dos Passos's prose denotes broadcast radio's spatial qualities, which were defined in the late 1920s by the emergence of afford-
able radio/loudspeaker sets that soon eliminated the solitary listening

of the crystal and headphone sets (although cheaper radios in the late 1940s would revive solitary listening [Crisell 1986, 11–12]). In tandem with the new listening arrangement, radios became increasingly associated with domestic space: the cabinetwork that housed their tubes and wires were increasingly designed to look like parlor furniture (as discussed in the first chapter, television would be "introduced" in much the same way), and advertisements for radio sets almost invariably placed them in the context of objects denoting the comfort and order of the bourgeois parlor (fig. 2). Unlike the telephone, which had an analogue in the telegraph, the radio entered the home without a technological analogue; because it is only a receiving medium that projects a transmission into ambient space, it does not so much extend self as it shapes the private space of the habitat, creating a sphere of radiance comparable to the circumambient heat of a fire—hence, there could be no more appropriate term than "Fireside Chat," a term that was shaped by both atavistic associations and the phenomenal characteristics of the medium (and Dos Passos gets at this through his setting— the house is "tiny" and "neat," full of "dryness, warmth, and light" while the storm rages outside). And thus, just as there was a mutation of domestic space when television was introduced, broadcast radio creates a heterotopia, a space of otherness within the habitat. The Fireside Chat was a revolutionary step in terms of the development of popular modernity's technophenomenal character, for it was capable of squeezing something as huge as national economic policy down to the cozy size of the typical American parlor, wrapped in the aural womb of "acoustic space" (cf. Carpenter and McLuhan 1960, 65–70).

It is also worth noting that this effect was not restricted by national or ideological lines, and it was refracted within the apparatus of cultural mediation and hypermediation in a number of ways. For example, the image of the national patriarch addressing his subjects in their cozy parlors through the medium of radio is found in popular paintings and photographs of the period: for example, in Germany, there is Paul Matthias Padua's *The Führer Speaks*, and in Italy, Richetti's *Listening to the Duce on the Radio*. Both of them remind us that the exploitation of media technology cut across the lines of national political ideology, for both of these paintings drew on cozy images of domesticity blended harmoniously with nationalist duty; in both paintings, we see the family patriarch seated in a position of power while the national patriarch, whose voice we assume to be pouring out of

Fig. 2

Introducing the Radio into Domestic Space. An Advertisement for a Radio Set,
1924, Smithsonian Institution American History Archives, Ayer Collection.

the loudspeaker, looks on from his portrait on the wall, thus reinforc-
ing the complicity between the concentric circles of patriarchy and
the use of technology to connect these circles (Mosse 1985, plate 17;
Pinkus 1995, plate 50). Richetti's painting was a prizewinner in a
1939 competition based on the theme of Mussolini and the radio. It is
worth noting here, as Cannistraro observes, that the Italian fascists
were somewhat slow in seeing the propaganda potential of radio, for
they didn't move to control it until the late 1920s. In spite of this, as
Pinkus notes, while the Italians focused more on the "pacifying or
relaxing" qualities of the radio as recreation for the middle and working classes, it is also the case that radio, as a domestic entity, was a form
of "forced camaraderie," and she quotes from one persuasional text
that advocates the abandoning of the practice of domestic privacy (husband smoking a pipe and reading in his study, wife and women friends
in another room, children in yet another room) in favor of the English-
style living room, and concludes that "a radio . . . cannot be situated
elsewhere than in the 'room for everyone,' of which it forms, as it
were, the nucleus" (qtd. in Pinkus 1995, 182). In this particular example, the propaganda text promotes the bourgeois ideal of domesticity
while the painting promotes its relationship to patriarchy and the state.
In all these mediations, radio technology is signified as the orderly
linkage between patriarchal domesticity and the larger claims of the
likewise patriarchal state (illustrated more forcefully in fig. 3). We see
that, as in the case of telephony, the use of popular modernity's technologies (technology within domestic space or with a phenomenal
"closeness" to the body of the subject) is both consumer-based and
ideologically forestructured through a variety of use-instructions.

A third area of significance regarding Dos Passos's "The Radio
Voice" returns us to the concept of hypermediation. While a cursory
reading of the passage would suggest only a kind of praise for the radio
voice, the security it can provide in times of crisis, and an approval of
FDR himself, there is here an undertone of caution. First, caution
regarding the content of radio programming, described here as "stale,"
"unfinished," "senseless," and, apparently, inductive of stupor or dis-
orientation (they listened "drowsily to disconnected voices"). The real
caution, however, is directed at the self-consciousness of FDR's radio
ethos ("painstakingly explaining how he's sitting at his desk there in
Washington . . . telling in carefully chosen words . . . so that youandme
shall completely understand that he sits at his desk there in Washington"),

Fig. 3

Radio and National Unity. "All Germany Listens to the Führer with the People's Radio," 1936, Bildarchiv Preußischer Kulturbesitz.

which could be taken as an index of the Fireside Chat's manipulative potential through its appeal to domestic sentimentality. Further, the phrase "not a sparrow falleth" is more revealing of this subtext, ironically announcing that the radio voice has transformed the elected public servant into an omnipotent god. Thus, Dos Passos identifies an undercurrent of demagoguery, a play on emotional insecurities, suggesting, perhaps, that if there is a link between European fascism and North American welfare-state democracy, then perhaps it is to be found concealed behind a softer sentiment, in a more discreet authoritarianism, and in terms of voice—in a more subtle intonation, but, in the final analysis, a similar deployment of the technologically mediated voice.

The passage, in short, hypermediates broadcast radio, but unlike journalistic mediations (which clearly praise or blame) or critical discourse (which often does much the same thing but with the praise/ blame function and its mythohistorical codes concealed by the alleged disinterestedness of academic inquiry) , it partakes in literary ambiguity and irony. Dos Passos had adopted a cautious attitude toward FDR when he first wrote the piece; but later, when he had more confidence in FDR, he tried to undermine his own irony by editing out the final paragraph when it was reprinted several months later in *In All Countries*. This editing, however, merely heightened the irony, for the phrasings discussed above remained, and eventually Dos Passos would side most decisively with his initial suspicions: in 1956 he would claim that FDR's approach to power "would have alarmed even the most authoritarian statesmen of our early history" (qtd. in Ludington 1980, 310, 414).

In short, Dos Passos ultimately agreed with Mussolini's initial estimate of FDR's political method—though they of course drew very different value judgments from this. This leads us to consider again the relationship between a media form and a political form: while there have been successful liberal radio personalities (such as Jim Hightower), for the most part, from the 1930s to the 1980s, from Coughlin to Rush Limbaugh, radio's political figures have been largely male conservatives, or perhaps reactionaries. Even Howard Stern, whose countercultural image and recourse to Sadean libertinism would seem to mark his kinship with either anarchism or the Left, is actually a neoconservative (conservative in fiscal, social welfare, and law enforcement; liberal on private behavior and sexual mores) who has thrown his support to the Republican Party and has voiced his support of capital punishment.

From Intimacy to Anonymity

While FDR was sitting at his desk in Washington perfecting the Fireside Chat, radio, the record industry, and the new technology of music was creating popular modernity's definitive entertainment "star,"
Bing Crosby. In the years prior to the emergence of commercial broadcast radio, phonographic recording began to replace sheet music as the
major form of popular music. The phonographic record marked a
transfer, first opened by the player piano, from popular music as a
participatory entertainment (and in that sense still closely affiliated with
the folk music from which it ultimately arose) to a more passively
consumed object. This entailed a major change: under the former
dispensation, the most important link between the popular music establishment and the consumer was musical notations (the "stardom"
of Stephen Foster, beginning in 1840s, established this earlier form of
pop music celebrity). In popular music after the eclipse of sheet music,
"the only text is the record. This makes the precise type and intonation of the singer very important" (Willis, qtd. in S. Jones 1992, 53).
The next important step was the introduction of the microphone,
which replaced the recording horn, which was nothing more than a
sheet-metal funnel around which musicians had to huddle in order to
transmit the vibrations from their instruments down the funnel, to the
needle, and finally onto the wax cylinder. The microphone and am-
plifier allowed for artificial control of the loudness and tone of the
singer's voice, and Crosby was perhaps the first and most successful
exploiter of the new medium. As Simon Frith notes, Crosby's new
style—crooning—was "made possible by the development of the microphone" (1986, 263), and indeed, crooning (derived in turn from
the Old English *cran,* "to moan") is a kind of singing that can only
proceed technologically, for its soft-voiced technique requires an amplification system. (This, incidentally, would not be the only instance
of Crosby's technological significance: he would also be the first pop
singer to make the move from direct recording on master disc to the
use of audiotape (S. Jones 1992, 39).

Crooning represented a new direction in music in that the personal qualities of the singer's voice could be emphasized just as power
and projection were de-emphasized, and ever since, popular singers
have been recognized by their particular vocal timbre or idiosyncra-

sies. (Mick Jagger's unique intonation, interestingly, resulted from the synthesis of a London dialect imitating black American/Southern speech with a barely perceptible lisp; the lisp came from an accident that caused him to bite off the tip of his tongue [Andersen 1993, caption to photo 4]). Crosby's crooning also was a perfect fit for radio's power to create the illusion of intimacy—whatever one might say about nineteenth-century popular opera or the popular music of the Victorian era, one cannot say that it promoted an "intimate" singing style. Rather, its approach based on projection capability, as still seen today in singers trained in either church singing (Aretha Franklin, Whitney Houston) or in the Broadway tradition (Ethel Merman, Barbra Streisand). Just as the parlor radio set facilitated an intimate atmosphere, as in FDR's Fireside Chats (we might call it "political crooning"), Crosby created an intimate bond with his female audience, and thus his kinship with latter-day pop stars is greater than might be suspected, for crooning is essentially a sexual discourse. The continuity of crooning is most evident in a subgenre of contemporary rhythm and blues—the seductive tradition established by Motown "crooners" like Smokey Robinson and continued in the more sexualized styles of Barry White, Teddy Pendergrass, and Luther Vandross. There is also a tradition of female crooners whose vocal styles, informed by the prenatal memory of the maternal voice as a "blanket of sound" (Silverman 1988, 72), serve the same fantasy role for the male audience; these styles parallel the range of sexual fantasy, from "sweet" (Diana Ross) to "earthy" (Tina Turner). In both the original crooning of the 1930s and in its derivatives, the voice is meant to suggest erotic power; as Crosby suggests in "Learn to Croon," a hit from the film *College Humor* (1933), crooning is the most reliable way to win the object of "your heart's desire." It would seem to be the case that the more mass-produced and technologically driven the communication became, the more domestic and intimate it was made to appear.

One of the Mack Sennett shorts of the period that features Crosby (*Sing, Bing, Sing,* 1931) can be summarized as follows: We see Crosby's sweetheart as she sits transfixed in front of a radio that emits Crosby's crooning. At the end of the song he broadcasts a private message just for her—to await his signal later that evening whereupon they will elope and escape her disapproving father and the undesirable suitor that her father has chosen for her (a plot from Roman comedy). Later, the father discovers the elopement plot and foils it; but the ever-resourceful crooner contacts the girl by phone to plot a second elopement. The crooner's use of both radio and telephone to invade the home for amorous purposes harkens back to the hysteria over the telephone "hello girls" alluded to earlier. Later, he uses a hidden phonograph playing one of his recordings to mislead the father and his henchmen, and ultimately, he and his intended make their escape in an airplane. Thus, the story evidences the intimate power of the crooner's voice, the crooner's use of radio as a medium, and the way in which radio and other media present new threats to paternal domesticity and habitat. This film, popular in its time as was Crosby, well illustrates my earlier point about the need to incorporate the role of hypermediation in an account of the social phenomenology of communications technology.

Crosby's crooning not only allowed a more sensual voice to "invade" the domestic sphere; it also allowed the racial Other to enter the domestic space of white America, for while from a contemporary perspective Bing Crosby's hits from the 1930s are thoroughly genteel, we may also hear in some of his work from the period the first faint traces of another voice, the voice of the American Other. In 1931 he recorded W. C. Handy's "St. Louis Blues" with the Duke Ellington Orchestra; in 1932, "My Honey's Arms," with the Four Mills Brothers; and throughout his career, as late as the mid-1950s, Crosby collaborated with jazz's great popularizer, Louis Armstrong. Pop/jazz singer Tony Bennett, in an interview regarding the role of Armstrong in American popular culture, instructs us that in Crosby's early scatting we hear, if we listen carefully, a sandblasted version of Armstrong's raspy scatting.

During the Jazz Age, bourgeois white Americans "discovered" the black American through a symbolic revolt against the puritanical and Victorian mores of their parents; in their desire to detach themselves from this "stifling heritage and its outmoded morality," they turned to African American music (jazz and blues) and dance (the Charleston) (Barlow 1990b, 175). This moral revolt, conceived as a form of transgression and resulting in a reencoding of imperialism in the form of cultural tourism, may also be viewed as an act of obedience rather than as a transgression (a "managed transgression," to be discussed in chapter 4), as the emerging consumer society required the abandonment of certain ideologies (e.g., that of thrift), and the construction of a black Other served as a facilitator for this transition. The black American, whose image had been controlled and used against him or her (particularly in the form of one of America's first popular entertainment forms, blackface minstrelsy) was also a part of the voice revolution. There was, first of all, a genuine interest in African American musical form and style, with its more complex approach to rhythm, its prioritization of improvisation and spontaneity, its reference to an African septonic scale, and its interest in timbral irregularity, expressed in "bent" notes and vocals characterized by roughness or raspiness. All of this, on a purely aesthetic level, was a welcome change from genteel Euro-American modes, to which jazz was in many ways diametrically opposed (European popular music prioritizes rhythmic regularity and a preference for "pure" tones). In Crosby, then, we find an early example of a very assimilated appropriation of black expressive form, a subject we will take up in much more detail in chapter 5.

But there is yet another voice that came to be. Whereas the voices thus far cataloged have a premodern origin—for authoritarianism (and relatedly, liberal demagoguery) are not inventions of the twentieth century, although modern forms of communications media, as indicated, have helped to expand their domains—the concept of "anonymous authority" does not. As Erich Fromm explains, the history of economics is one of ever greater impersonality and instrumentality as the marketplace transformed from a physical meeting place to an abstract mechanism that works to establish commodity exchange values on the basis of supply:demand ratio. The market concept of value (which prioritizes exchange over use), through the historical process of ideology formation, came to have a marked effect on the modern concept of self, for there is no cordon sanitaire between economics and ideologies of the self. Fromm describes this as a

> character orientation which is rooted in the experience of oneself as a commodity and of one's value as an exchange valued. . . . In our time [this] marketing orientation has been growing rapidly, together with the development of . . . the "personality market." Clerks and salesmen, business executives and doctors, lawyers and artists all appear on this market. It is true that their legal status and economic positions are different: some are independent, charging for their services; others are employed, receiving salaries. But all are dependent for their material success on a personal acceptance by those who need their services or who employ them. (1947, 68–69)

The promotion of "personality" as an item with exchange value has led to the development of an interchangeable public personality whose function is that of exploiting and facilitating the market. As "personality" (i.e., "uniqueness") has come to play an increasing role in commodity exchange, it becomes commodified, and the key to commodification in a mass market is interchangeability (74–75). Ironically, then, the economic valorization of the individual, unique personality (if indeed there ever was such a thing) has led to the interchangeable (i.e., nonindividual, nonunique) "personality." And this has repercussions in the realm of voice: "the voices of speakers," Theodor Adorno noted, "are meeting the same fate as befell, according to psychology, that of conscience, from whose resonance all speech lives: they are being replaced, even in their finest intonations, by a socially prepared mechanism" ([1951] 1978, 137).

An illustration of this can be found by returning to our discussion of the telephone. Nothing could better illustrate the thesis laid out here by Fromm and Adorno than the social history of the telephone. As Michelle Martin (1991) notes, with the increased use of the telephone, the disembodied voice became a "long distance agent of sociability," and thus it became "increasingly perceived as revealing the personality," which in turn led to the notion that it must be educated through voice training. In the hands of the emerging telephone companies, the human voice was something that had to be standardized. There are a number of determinants in this process, the first of which is social class: as Martin notes, "[C]ourtesy was associated with bourgeois and upper middle-class manners. A working-class voice could not be courteous if it was not trained to be 'lady like.' A rough or high-pitched voice could not belong to someone from the affluent classes" (1991, 91), and this was compounded by the distaste for electronically reproduced female voices, as discussed by Amy Lawrence (1991, 31). Furthermore, standardized practices would save response time (by insisting on stock phrases like "number, please") and increase the rate of production by decreasing personal contact with the customer. Perhaps most importantly, standardized voices would serve to avoid the "perception of transience" for the telephone company, for no matter how often operators might be transferred, quit, or be fired, there is an illusion of permanence in the standardized operator voice, as though it were the same operator every time (Martin 1991, 94). Further, the attempt to control the female voice of the operator reveals a deeply structured gender conflict present in the sociology of technological adaptation, for the telephone company worked hard in the early decades to promote the business uses of the telephone (male) and to suppress the social uses (largely identified as female).

The voice of the operator is reflected in a host of other anonymous voices of popular modernity: the invisible narrators of public life that follow us through the day in the admonitory voice of radio and television commercials, the underside of which is the private (rather than public), anonymous voice of phone sex, which Stone calls "the process of constructing desire through a single mode of communication, the human voice" where, loosened from their anchors in physical experience, "the most powerful attractor becomes the client's idealized fantasy" (1995, 94). Anonymous female voices and the voices of female popular singers counterpose the voice of the father, for such voices can be both "an acoustic reflection of male desire" and of nostalgia for the maternal voice (Miller Frank 1995, 37). Thus, technologically transmitted anonymous voices (and the voices that never were voices, in the case of computer generated speech, an increasing feature of software programs) frame the world of popular modernity—from the mundaneness of K-Mart (with its voices announcing the "flashing blue light special") to the underground (yet popular) world of pornographic self-stimulation.

Hypermediating the Radio Voice

Earlier in this chapter, we considered a journalistic specimen regarding the role of wireless telegraphy in the attempt to rescue the passengers of the *Titanic*. This particular text, and many others like it, conveyed the sense that wireless broadcasting was part of a sublime technological frontier, the contemplation of which was as entrancing as was the prospect of ongoing American technological superiority. In the early 1920s, print media, and newspapers in particular, glorified radio in the hope that the two media could exist symbiotically and that ultimately the attention paid to the radio phenomenon in newspaper articles would increase circulation (Patnode 1998). In this way, Susan Douglas argues, newspapers contributed to the social construction of broadcasting, and we should recall here similar transmedia alignments, discussed in the previous chapter, between print journalism and the telegraph, and later, the Internet. As Douglas demonstrates, journalism "constructed" broadcasting largely through narratives that, contrary to the technology's twenty-year history involving private inventors, presented radio technology as a solely corporate affair and a necessary part of the American home (1987, 292, 304). By the latter 1920s, however, the newspapers tended to view radio as a threat; particularly threatening was radio's ever-expanding role in providing news coverage. This culminated in the 1933 "Press-Radio War," which was settled by the Biltmore Agreement, which required radio news to rely exclusively on the Press-Radio Bureau (cf. Jackaway 1995).

The attitudinal stance of both the literati and the journalism establishment toward broadcasting that developed in the 1930s is demonstrated in the following passage:

> One of the great problems before our civilization today is the sinister insidiousness of radio. Long after Coughlin has passed out of the picture . . . the insidious radio will still be before the people. . . . It steals into the home with its whispered words, coming from no man knows where. It is a voice and it is gone. There is no record. The is no permanent printed word. The poison of the demagogue, of the atheist, the communist, and the lecherous fills the air of the home and is gone, leaving its stain. Vile and suggestive song, words of double meanings, pour fourth to be subconsciously accepted. And there is no written record to prove the injury, no way of combating the evil that is done. (Qtd. in Warren 1996, 51)

This passage by columnist Malcolm Bingay, which appeared in the *Free Press* in 1938, suggests that the character of radio's hypermediation changed considerably of the course of twenty-five years (the earlier *Titanic* passage had been written ten years before commercial broadcast radio; the Bingay passage, about ten years after radio had been firmly domesticated).

The *Titanic* and the Bingay passages suggest, among other things, two groundings of American ideology—the puritanical and the romantic—and their two corresponding genres, the jeremiad and the lyric. In this case, we find that the romantic precedes the puritanical (evidently these ideological groundings are cyclical rather than linear-progressive). Further, while I do not wish to imply that most people, in their subjective day-to-day experience and use-interactions with technology, were burdened by either undue unconscious anxieties of this sort nor even conscious concerns, the presence of this hypermediation does indicate that these anxieties were present in the intersubjective, intertextual web of culture.

Finally, Bingay's text needs to be contextualized with both the Dos Passos and the Innis texts, for even though all three bear the specific markers of their generic practices (respectively, the reductive vitriol of the journalistic opinion piece, the ambiguity of the literary text, the analytic matrix of the critical text), they all nonetheless perform an essentially similar hypermediation in that they present, to one extent or another, a hostile mythologization toward an Other (nonliterary) media and postulate a postliterate assault on culture. The Bingay passage is furthermore instructive in that it conjoins radio's full range of practices—from the crooner ("vile and suggestive song") to Hitler and Coughlin ("the demagogue"), from the excessively authoritarian to the sexually transgressive—for the purpose of mythologically demonizing the aural and psychological characteristics of the medium rather than any particular discursive practice per se.

The fascist ruler, the charismatic leader of a representative democracy, the pop singer/idol, the appropriated voice of the racial Other, the anonymous voice of the commercial announcer or the phone sex operator: these are the disembodied voices of popular modernity. And while it would be epistemologically desirable to come to some final stance regarding the essential phenomenal character of these voices, it seems we must settle for a mere charting of their shifting identities in terms of the way a medium has itself been mediated and the way experience has been mythohistorically forestructured.

3

Temporality and Commercial Culture: Nostalgic Entertainments

Time, Culture, Phenomenology

In our consideration of the phenomenology of listening and voice in the case of broadcast radio, there was a very basic element that we did not adequately acknowledge, which is simply this: the sound of the voice "unfolds" over the course of time. Indeed, in most of our deliberations thus far, deliberations concerned with spatiality as well as listening and voice, we have approached phenomena in a more "punctal" than chronodynamic manner. Contrary to this, however, all forms of experience (such as reading a map, for example) are implicated in diachronic complexities.

We will have occasion in the closing chapter to address the phenomenology of time specifically as it relates to narrative. In this present chapter, however, I will confine the discussion to the problem of time as it relates to the experience:culture problematic as a prelude to a discussion of what Stephen Carter calls "American time," which he sees as being rooted in both consumer culture and a kind of cultural "chronophobia." My focus here, unlike previous chapters, will be much more on specific entertainment forms (popular music and film, with particular reference to the nostalgic mode) in the period after which the technologies and distribution systems that support modern entertainment had been firmly established and fully naturalized; in other words, in the era of a fully established popular modernity.

* * *

A phenomenological reduction of time perception is difficult to accomplish for a number of reasons. For one thing, time is not necessarily associated with any given sensory mode. Of course we listen with our ears and see with our eyes, although even here we need acknowledge that the experience of spatiality, as I've argued, is something beyond mere seeing. But time is even more problematic: How do we sense its passing? Do we "watch" it pass? "Hear" it pass? Do we simply "feel" the passage of time somatically, cyclically through the "biological clock" of repetitive cellular processes and linearly through the aging process? The sense of time's passing (and the fact that the acculturated nature of these descriptions is revealed in the very choice of the word "passing"—as opposed, say, to "marching" or "flowing," or, as I said before, "unfolding") according to Husserl (following Brentano's example) can only be clarified by first coming to terms with a fundamental paradox peculiar to time consciousness. That is, we are apparently able to perceive, simultaneously, that which is present and that which is not, a condition that allows us to hear and perceive not merely tone, but melody as well:

> While the first tone is sounding, the second comes, then the third, and so on. Must we not say that when the second tone sounds I hear *it*, but I no longer hear the first, and so on? In truth, therefore, I do not hear the melody but only the particular tone which is actually present. (Husserl [1928] 1950, 43)

Husserl brings closure to this paradox by positing that every "now point" in time consciousness is of *"what has just been* and not mere consciousness of the now-point of the objective thing appearing as having duration" (54), and he articulates this concept in terms of a complex dynamism, employing a number of descriptive markers that were ultimately incorporated into the analytic approaches (in phenomenology, hermeneutics, and aesthetic theory) of Heidegger, Schutz, Merleau-Ponty, Ingarden, Gadamer, Iser, Jauss, and others—concepts such as "retention," "protention," and "horizon" (retention being the echo of the moment that has just passed through consciousness and thus the past inscribed in now-consciousness; protention, an expectation of the probable future as it exists in now-consciousness).

Time perception is thus, according to Husserl, a function of a dialectic of ever-transforming memories and modified expectations, both of which he metaphorically renders as the horizons of a spatial field, which transform and in turn are transformed by the now-point. Thus, at any particular moment of our listening, a melody consists of a layered perception—the simultaneous awareness of the tone that presents itself to us physically through our ears, those tones whose moment has just passed, and some sense of the melody's telos.

Listening, as Husserl's account demonstrates, is far more complex than it appears; and the complexities he identifies are all the greater when cultural considerations are brought into the formulation. Music provides a good example here. Brentano's—and later, Husserl's—selection of music as a stock example of temporal problems in phenomenology is by no means arbitrary. We noted earlier that our experience of spatiality is something we generally regard as being acultural, and it is much the same with our attitudinal stance toward instrumental music. More that any other art form, instrumental music presents itself to us as nonreferential. As Walser remarks, "[W]e experience music's rhetorical pull apart from language, seemingly apart from all social referents, in what is usually thought a pure . . . way" (1993, 113). Thus it accommodates the tendency in traditional phenomenology (and in Kohler's Gestalt psychology) to deal with the world in terms of primary perceptions—shapes, colors, tones. But this idea about music—that it is nonreferential and provides an unmediated experience—is itself just that: an idea, a Romantic idea at that, and thus generally accompanied by a heroic rendering of the musical artist, a rendering we see in Nietzsche's Dionysian musician, who is "without any images, himself pure primordial pain and its primordial re-echoing" ([1872] 1967, 50). We can see that the phrase "pure primordial pain" and others like it have an acculturated dimension, and the very fact of this acculturation undercuts the purport of such phrasings. And it could not be otherwise, for again, the idea that music is beyond culture is itself a cultural idea.

And thus, the listener may indeed perceive melody in the manner described by Husserl—but the model must be semiotically problematized if it is to be a more accurate description of actual perceptions in actual cultural and historical circumstances. If the melody is, say, from Mozart, will not the listener's phenomenal frame be modified by whatever it might be that the sign "Mozart" signifies for him or her? And, given any one individual's class and cultural parameters, the nature of this semiophenomenal nexus can vary greatly—and this is so even if we assume equal levels of familiarity and competence. What about timbre? For example, a melody played on the violin will have not only a timbral difference from one played on the trumpet but a semiotic one as well, for the violin and its tone, for the Western ear at any rate, signifies warmth, emotion, and the individual soul, while the trumpet and brass in general often have atavistic military associations. The feelings we associate with various instruments may have little to do with the qualities of the instrument itself and may in fact be largely the result of semiotic reference systems that are "always already" forestructured. This postulate becomes more acceptable if one has had the experience of initially encountering a given musical form (particularly a foreign one) as "noise," only to later develop an appreciation for it—not so much through mere repetition, but through the unconscious accumulation of cultural competence in the form of evaluative and aesthetic codes that are the very atmosphere—invisible yet essential—in which specific art forms generate and evolve. And indeed, the trumpet/violin example is merely an example of what Attali identifies as the history and the "political economy of music"— a "succession of *orders* (in other words, differences) done violence by *noises* (in other words, the calling into question of differences) that are *prophetic* because they create new orders" (1989, 19). To draw another example: suppose there were two renderings of a selection from Wagner's *Das Rheingold*, the first done with the usual instrumentation and the second with harmonicas and kazoos. A Trobriand islander might not perceive that the first rendering is "serious" and the second parodic. But we would, for we interpret according to a binary coding that associates the violin's timbre with "seriousness" and that of the kazoo with "nonseriousness." "We may say," Robert Walser muses in his Bakhtinian probing of genre and discourse in popular music, "that a C major chord has no intrinsic meaning; rather, it can signify in different ways in different discourses, where it is contextualized by other signifiers, its own history as a signifier, and the social activities in which the discourse participates" (1993, 27; cf. 46–48).

This considerably problematizes classical Kantian and Husserlian approaches to aesthetic experience, for such approaches seek to render these experiences as independent of culture. One response would be to suggest that there is, at the very least, a form of aesthetic response that is relatively independent from the dictates of cultural codes. One thinks of the Kantian distinction between delineation and color in which color is related to the merely pleasant, whereas delineation and its attendant complexities are the element that triggers the aesthetic judgment proper ([1790] 1914, 73). One could construct a similar bifurcation between culturally-bound and nonbound aesthetic experiences that would be more purely perceptual in nature by virtue of their correlation to universal experiential categories, but I think this would be misleading. First of all, the most recondite facets of perception are so close, so "natural" to us, that their "acculturatedness" is difficult, perhaps ultimately impossible, to demystify. To put it another way, Kant's universal categories of time and space may be, as Kant himself suggested, like a pair of invisible spectacles that we cannot remove. But it is also true that there are many different kinds of spectacles that refract space and time in different ways. And so, not only are specific aesthetic judgments and their criteria culturally coded (judgments regarding, for example, a Wagner score performed on a kazoo)—but so too are the perceptual categories upon which human experience, including aesthetic judgment and experience, are based.

Time, throughout most of time, was reckoned to be essentially cyclical, following the recurring events of the natural world—birth, growth, maturation, death, and a decay that leads once again to birth. Also, to return to Benedict Anderson's analysis of the prenational "transcontinental sodalities" of Christendom and Islam, time had not only an agricultural cyclicity but a mythic depth and a transcendent relationship with the believing subject. In this concept of time, "cosmology and history were indistinguishable, the origins of the world and of men essentially identical," and this, along with other factors in the preindustrial, prenational cultural system, provided a certitude of meaning to the "everyday fatalities of existence (above all death, loss, and servitude) and offer[ed], in various ways, redemption from them" ([1991] 1994, 89). (As we noted previously, however, such historical judgments are themselves enactments of myth, as in this reenactment of the myth of the Fall.)

Under industrial capitalism (in either fact or myth—perhaps both), then, this *Zeitweltanschauung* was significantly reencoded: whereas in earlier epochs time was essentially passive (something that simply "passed by"), it now became "a resource and a basic dimension of societal-economic livelihood and prosperity," as Ahron Kellerman puts it (1989, 39), and thus transportation and telecommunications technologies were used to find new ways of mining this resource. As noted in our discussion of spatiality, this revolutionary change in communication technology informed, in the second half of the nineteenth century, the economic shift from arbitrage (speculation based on spatially separate regional markets) to futures (in which space collapses and speculation is based instead on time—the possible future value of a commodity). As Lyotard remarks, this form of economy requires a habit of mind in which the future is taken as if it were present (qtd. in Carter 1994, 38). The future-orientation of late-nineteenth-century capitalism has taken on an enhanced quality in the United States, for in the period between the War of 1812 and the Civil War, owing to the need to create a national identity separate from England and to find cultural amelioration for the relative absence of a national past, the young American republic already had a future orientation. In fact it had already been presaged by the Puritan typological method of biblical interpretation and, more to our purposes here, ideologically codified by an emergent American intelligentsia (e.g., Emerson, Whitman, Fuller). In toto, the imperative is to exploit the present and the future, for to do such was not only the directive of the marketplace: it was a cultural desideratum produced by the American perception (and my use of the term "perception" here is a reminder of the very culture:experience problematic I'm getting at) that since an adequate claim on the past could not be staked, perhaps an alternate claim could be laid on the future.

Popular Nostalgia

The contemporary American ideology of time can be uncovered by tracing the history of nostalgia. The term, Fred Davis notes (1979), was coined by a Swiss physician, Johannes Hofer, in 1688; Hofer had apparently succumbed to the classical fashion of his time when he discarded perfectly workable vernacular terms (the German *Heimweh,* similar to the English "homesickness") in favor of the Greek *nostos* (to return home) + *algia* (a painful condition). Davis goes on to point out that this term, originally a medical one (the theory was advanced by another physician that the Swiss were particularly prone to this because of the change in atmospheric pressure experienced when one abandons the mountains for the lowlands), later became a military term (homesickness became particularly evident in the massive armies that were formed by the "universal conscription" policies of Napoleonic and post-Napoleonic Europe), and later still became a psychological term.

In the post–World War II era, however, the sign "nostalgia" became reencoded as a result of the shifting ideology of time; from sometime in the 1950s on, all of these former usages became "dissipated through . . . popular and commercial usage" (F. Davis 1979, 5). More importantly, for baby boomers and beyond, nostalgia is not a longing for place; it is a sentimental and generally distorted longing for the past (for the "simpler days gone by" as presented to us by both myth and what purports to be history but which is actually mythohistory) that is encouraged and exploited with great energy by the entertainment industries. This raises the question: how is it that a society that for more than a century was primarily future-oriented became so given to an interminable probing of the recent past? There are a number of tenable explanations.

First, in the postindustrial phase, the era of consumer capitalism, the subject continues to be conditioned by the "future orientation," but with a difference. The future orientation, a product largely of the era of producer capitalism, is now, paradoxically, part of the past, although it is somewhat successfully buoyed by a mythohistorical orientation that has been transmitted through public education, political rhetoric, and popular narrative. The cultural contradiction here—the idolization of a future orientation because that was the way America was successful in the past—oddly conflates the two directions in time. Further, the economic and technological disruptions of modernity, taken together with a plethora of messages regarding the disruptions of modernity, make it all the more easy to market the past as a mythic home, a place of peace and stability. Second, one might turn to the evolution of material culture. Throughout history material objects have been fetishized and used as focalizers for the purpose of conjuring a sense of connection with the past (for instance, with one's family ancestors), and as consumerist culture provides us with a welter of personal objects, there is plenty of opportunity to continue these practices.

One modern fetish object that is of supreme significance is the photograph. We shall return to the subject of photography in the next chapter, which is devoted to the phenomenology of the popular image, but here we must make some remarks regarding photography as it related to temporal perception.

The commercial photography studio established itself as a commonly available service in the larger cities of the eastern United States and Canada by the mid-1850s, catering heavily to the family portraiture market, and family photographs became all the more available, and eventually ubiquitous, with George Eastman's invention of the "Kodak" in 1888. We have noted the introduction of technologies into the domestic sphere earlier, and we see the same process here again, for as Hirsch notes, "With the slogan [Eastman's] 'you push the button, we do the rest,' the camera entered the domain of the ordinary and domestic" (1997, 6). With the Kodak camera widely available, and with its uses being directed (hypermediated) by advertisements that advocated the photo collection as an essential element of normal American domesticity, there occurred a technological reencoding of time and memory, for "photography quickly became the family's primary instrument of self-knowledge and representation—the means by which family memory would be continue and perpetrated, by which the family's story would henceforth be told" (Hirsch 1997, 6–7).

Arguably, with a change in the technologies of personal and familial memory there comes a change in the character of these memories. As Russell Belk notes, the "snapshots that fill our drawers, slide trays, and family albums do not present an honest portrait of our everyday lives." Photographs thus comprise a "heavily biased" selection, a personally-directed mediation of the moments of our lives—"selective repositories that portray the selves we wish to preserve for the future" (1991, 114). The way in which photography can selectively direct memory has larger historical, political, and cultural implications: we may think of the general understanding among news photographers during the Roosevelt years that the president was not to be photographed on his crutches. A more sobering example is found in the "crucial moment in Czech history" that serves as a point of departure for Milan Kundera's composite novel, *The Book of Laughter and Forgetting*: a 1948 photograph of ranking Communists, including Party leader Klement Gottwald, was later altered to remove the image of Foreign Minister Clementis, who had fallen from grace and thus needed to be "airbrushed out of history" (Kundera 1980, 3). Those who were there, however, would remember that the hat on Gottwald's head was actually Clementis's (Clementis had, in an act of deference, placed it there to protect the leader's head from the snow flurries). Kundera uses this act of deception to introduce the theme of historical amnesia and the "struggle of memory against forgetting" that informs the various stories in the novel.

In the emergence of commercial photography, we see a transformation of nostalgia—from a longing for place to the revision and commodification of the personal past, and we can postulate that the public's acculturated taste for sentimentalized autobiography shades over into other consumer habits; this is particularly true of popular music. On the level of personal time, music often causes one to "flash back," in Bergsonian fashion, to "where we were" when we first heard it, and we may drift into reverie or experience nostalgic longing. Popular music in particular is experienced in this chrono-indexical mode in which the phenomenology of primary aesthetic reception is supplemented and supplanted by the music's association with fragments from one's personal past. It first must be said, however, this is not the only way in which popular music is consumed. Another related set of questions regarding perceptual awareness that is largely untouched by Husserlian analysis has to do with retentions that go beyond any discrete spatiotemporal phenomenal event. For instance, music may speak to us from what Schutz calls "the world of predecessors," thus serving not merely the purposes of personal memory but that of cultural memory as well. This is suggested by James Baldwin's short story, "Sonny's Blues," particularly at the end, when the older "mentor" musician guides the young protagonist on a musical journey back from bebop jazz to the blues. This conceptualization of popular music is reflected in George Lipsitz's focus on "collective memory" in his study, a notion well summarized by David Bowie: "In our music," says Bowie, the blues are our mentor, our godfather, everything. We'll never lose that, however diversified and modernistic and cliché ridden . . . it becomes. We'll never . . . be able to renounce the initial heritage" (qtd. in Lipsitz 1990, 103).

Having said this, however, one needn't search long to find evidence of the manner in which the nostalgic is promoted, and that this nostalgic attitude probably governs much of how the music is used (although again, various forms of use can coexist within a single human subject and even within a single perceptual act). For instance, a 1996 television commercial for a Time/Life music CD collection called *The Heart of Rock and Roll Series* begins with the narrator asking, "[D]id you ever fall in love so hard that it hurt?" If so, you may "relive these magical moments" by listening to this collection, which is organized chronologically, perhaps so that it can all the better parallel the listener's self-narratization. It is only later that the commercial's narrator tells us that we can "recapture the music," thus indicating that the music itself is to be regarded as a facilitator for self-referential pleasure (or pain, as the conjoining of "falling in love" and "hurt" implies). More perplexing chrono-ironies surface when we are encouraged, at the end of the commercial, to charge the entire collection to a credit card: we are given the option of spending money that we don't presently have (but which we can pay off in the future, at the price of additional interest) in order to buy a commodity that purports to bring us back to the past. And so we phone in our credit card order in the belief that, in the near future (when the CD set arrives in the mail), we will receive a product that brings us into the past at the cost of funds we hope to have in the future. The tangle of technologies (television advertising, computer-enabled instantaneous credit, enhanced digital re-recording) and temporal relations (imagined mythic past, posited financial future) suggested by this example is a structural homology to the cultural and economic reversal that takes place in the shift from the first stage of industrialism to Fordism (populism, consumerism, consumer credit, etc.), for in the former the saved resources of the past are used to build a future, while in the latter the future is used to purchase the past.

It is in this manner that for the past two decades, one of the most conspicuous entertainment trends has been that our culture has obsessively "demanded and produced countless tributes to our public art of the last fifty years" (Graham 1984, 351). Or as Sven Birkerts puts it, "[O]ur culture is awash like never before in repackaged bits of the past" (1989, 20). Moreover, given the ongoing competition among those popular culture industries that serve this nostalgia market (and in so doing progressively saturate that market), given that this past is limited to that which is within living memory, and given that much of the consumer market is oriented toward persons under twenty-five years of age (whose sense of what constitutes "past" will differ from that of older people)—given all these things, there has been a tendency to redefine "past" in such a way that the boundary between past and present has become increasingly attenuated.

We can witness this process of attenuation if we first consider the nostalgia represented in the Hollywood musical during its heyday from 1946 to 1962. Several musicals from the period, paralleling and influencing the dominance of the television Western genre, such as *Annie*

Get Your Gun (1950) and *Calamity Jane* (1953), reference the mid- to late-nineteenth-century American frontier (thus once again indexing Turner's great divide), while others, those belonging to the subgenre of the "fairy tale musical," reach back before the time of Manifest Destiny, reversing its east to west directional narrative, to place the spectator in an imaginary premodern world. Thus, *Brigadoon* (1954) brings two unhappy New York moderns (one a romantic; the other an alcoholic) to Scotland, where they enter a magic, invisible village frozen in the eighteenth century (a fairly atypical referencing), thus providing a fantasy escape from the "perversities of modernism" (Altman 1987, 170). Other musicals from the period, such as *Meet Me in Saint Louis* (1944) and *Take Me Out to the Ball Game* (1949) reference a nearer chronological frame—that of an imagined prelapsarian America at the turn of the century. In these examples, the chronological distance between the date of the nostalgia product's inception and the referenced time frame is forty to sixty years, and in the case of the fairy tale musical, considerably more. The nostalgic referencing of the film musical follows this matrix in a fairly predictable way, although one film, *Singin' in the Rain* (1952), marks the self-referential postmodern shift by referencing the film industry's own past in the cataclysmic technoshift from silent to sound in the early 1930s. While the film musical is notoriously self-referential (they are shows about people putting on shows, as in the exemplar of this plot—Mickey Rooney and Judy Garland putting on a show to help orphans in *Babes on Broadway* [1941]), the performances referenced are invariably traditional live stage performances. As Feuer notes, "[T]he Hollywood Musical worships live entertainment because live forms seem to speak more directly to the spectator" (1993, 23). *Singin' in the Rain*, on the other hand, takes this self-referentiality further by being a film musical about film musicals. This film's frame of nostalgic references is only about twenty-three years, which is fairly atypical given its date of production: the time referencing of nostalgia-based musicals from the same period is much longer, and it is not until much later, with *Grease* (1978), that we see evidence of the new "foreshortened" nostalgia (table 1).

By the early 1970s the nostalgia market was typified by this foreshortening; from that time on, nostalgia products typically have a referenced time frame of about fifteen to twenty-five years in the past. The rock band *Sha-Na-Na* introduced 1950s nostalgia at the Woodstock Festival (1969), with more mainstream products, such as the film

Table 1. The Incremental Foreshortening of Nostalgic
Referencing in Representative Hollywood Musicals, 1944–1978

A. Film	B. Year Produced	C. Frame (Setting)	D. Span (B minus C)
Meet Me in St. Louis	1944	1903	41
The Harvey Girls	1946	1880s	50
Centennial Summer	1946	1876	100
Three Little Girls in Blue	1946	1902	42
The Shocking Miss Pilgrim	1947	1847	100
Mother Wore Tights	1947	1900	47
The Pirate	1947	early 1800s	130
Easter Parade	1947	1912	35
Take Me Out to the Ball Game	1947	ca. 1900	47
In the Good Old Summertime	1949	ca. 1900	49
Showboat	1951	mid–1800s	90
Singin' in the Rain	1952	ca. 1932	20
Brigadoon	1954	1754	200
The Music Man	1962	1912	50
Grease	1978	ca. 1960	18

American Graffiti (1973) and its television spin-off, *Happy Days* (1974–84), referencing the same time period. By 1995 this reference point would remain the same, at about a distance of twenty years (i.e., 1975, as demonstrated by the 1995 release of *The Brady Bunch Movie*) but with the addition of a nostalgic approach to the 1980s—as in yet another Time/Life Music advertisement (this time in a mailing circular distributed in 1996) promoting a CD collection called *Sounds of the Eighties* ("The '80s were just too much fun. Why shouldn't you be able to bring it all back?"). MTV has taken a prominent role in this new, more radical form of foreshortened nostalgia, as in a 1995 production entitled *It Came from the Eighties* and a 1996 production entitled *The Nineties Up to Now.* (Showing up at the same time was a commercial for yet another CD set called *Living in the Nineties* and a book venture not affiliated with MTV but sharing its sensibility, *Glossary for the 1990s.*) The titles of these productions demonstrate the now-customary habit of thinking of the past in convenient, take-home packages of ten years to the bundle—even if there are more historically justifiable bundles (i.e., popular music from 1954 to 1963, a period that delineates a period of formation, establishment, and degeneration for the first phase of rock and that, perhaps not coincidentally, parallels the civil rights movement from *Brown v. Board of Education* to the passage of federal legislation). But with the marketing of 1990s nostalgia in 1996, "decade nostalgia" can be marketed even if the pack is short four or five years. The use of the terms "Living" and "Now" in the titles for these new nostalgia produces indicate that the logical conclusion to all this would be the development of a nostalgia, not only for last year or the year before but for last month, yesterday, or perhaps a nostalgia for the present itself, as the commodity becomes not an artifact from the past itself but merely the sense of dissociation from it. A recent, techno-necrophilic twist on this is the trend of recording music with the dead, as in Natalie Cole's duets with her father, Nat "King" Cole, or the 1995 recording of "Free as a Bird," a collaboration between the three living Beatles and a 1977 recording of a solo performance by slain Beatle John Lennon. The "Free as a Bird" recording and its temporal peculiarity became a leading element in a major Beatles nostalgia campaign (interesting in itself, for the Beatlemania of the early 1960s was rooted in a thorough rejection of the past in order to embrace the new pop aesthetic, which, ironically, was largely a repackaging of black American music).

Steven Carter postulates a cause for these cultural phenomena by grounding human temporal experience in the two opposed conditions—being "with" time and being "in" time:

> [W]hen a human being or an entire culture refuses to accept [time] . . . and desires to escape from it, then they fall into a state . . . which might be termed "withness." Withness denotes an ontological separation: one is not felt to exist in time, or to be contained by time, as a glass contains water or as water contains fish; nor is one's experience felt to be constituted of time, as water is constituted of hydrogen and oxygen atoms. On the contrary, the desire to escape from time automatically makes of time a separate entity detachable from the self, a "thing" which always exists somewhere or "somewhen" else. (1994, 37)

This "withness" of American time is particularly apparent in two popular genres that Jameson identifies as being related "through a relationship of kinship and inversion" and that explicitly addresses temporal orientation; these two genres are the nostalgic narrative and the futurologic narrative (1989, 523). The accuracy of Jameson's formulation is borne out by a plethora of Hollywood films that are both—we may call them futurologic-nostalgic narratives.

As with the phenomenon of nostalgia, the science fiction genre has its origin in an ideological and experiential shift concerning the concept of time. Mythologization of the future, as evident in the myth of the Norse Ragnarok—the destruction of the world at the end of time—takes on a particular character in popular literature of the industrial age, beginning with the technofantasies of Jules Verne. As Scholes and Rabkin put it, the science fiction genre and its development reflect "the history of humanity's changing attitudes towards space and time" (1977, 3). They furthermore posit the genre's origin in the prior discourse of realist fiction, which developed as a rejection of myth in favor of a rationalist/empiricist discourse practiced as a normal science preeminently oriented towards the description of the present, as in the major novels of Balzac and Zola. Predicated on this earlier development, science fiction emerged as a rationalist/speculative (as opposed to rationalist/realist) discourse practiced, at least in some sense, as a revolutionary (rather than normal) science preeminently oriented toward a description of the possible future and occupying the generic space left by a utopian literary tradition that had focused on heterotopia (otherness of place), which would henceforth be, under the aegis of science fiction, primarily concerned with the description of the heterochronos: otherness of time; specifically, the future.

This, at least, would be a fair way to describe the central line of literary science fiction from Shelly, to Bellamy, to Wells, and even to the dystopians Zamyatin, Huxley, and Orwell. The speculative element of classic science fiction was first recognized by Percy Shelly in the preface to Mary Shelly's *Frankenstein* (1818), in which he claims that the central event of that novel, while impossible, created a hypothetical reality with an ability for "delineating human passions" that goes far beyond plausible fictions (qtd. in Alkon 1994, 5). Wells extended this view considerably when he declared, in a lecture delivered in 1902, that there is a "distinction between . . . the legal (past-regarding) and the creative (future-regarding) minds" and that science could go very far indeed toward predicting the conditions of the future (qtd. in Scholes and Rabkin 1977, 15–16).

While the technophobic narratives of the 1950s science fiction film do to some extent conform to the "what if" principle of speculative discourse (some far more so than others), they are generally far more invested in the aesthetics of horror and a reduction of Victor Frankenstein's Faustian complexity to a caricature—that of the "mad scientist." Another common transformation, and this is what specifically concerns us here, is the amalgamation of science fiction with the American mythic narrative, specifically the western and its spawn, the action-adventure genre. This move can be detected as early as the *Flash Gordon* series (1936–40), but it flourished with particular energy in the *Star Trek* television series (1966–69, 1973–75), which occupied the vacuum left by the gradual disappearance of the television Western. The uniquely successful *Star Trek* formula combined the Turner frontier mythohistory of the Western genre with the speculative orientation of classic science fiction. As Slotkin puts it, "[T]he displacement of the Western from its place on the genre map did not entail the disappearance of those underlying structures of myth and ideology that had given the genre its cultural force. Rather, those structures were abstracted . . . and parceled out" (1992, 633).

The 1977 film *Star Wars*, which received some of its impetus from the ongoing success and cult status of *Star Trek*, begins with the phrase, "a long time ago, in a galaxy far away," thus placing the "future" in the past, and thus we have arrived at the nostalgic-futurologic, a modality that is semiotically reenforced throughout the film in objects that are both archaic and futuristic, as in, for example, laser weapons that are wielded in combat like heavy medieval broadswords. In early 1997, following the now-standard twenty-year commercial nostalgia curve, an enhanced version of *Star Wars* was once again the top grossing movie in the United States, and it had loads of peripheral support (a Taco Bell marketing tie-in; a two-billion-dollar marketing agreement with Pepsi; plastic toys, etc.). In 1999, some more interesting ripples in the temporality-entertainment nexus were created with the release of *Star Wars: Episode I—The Phantom Menace,* which, in terms of plot, precedes the 1977 *Star Wars,* while at the same time, Lukas demonstrated his love for the future by announcing the proposed release dates for both *Star Wars: Episode II* (2002) and *Star Wars: Episode III* (2005) (IMDB 1999).

When one inquires as to the sudden reawakening of consumer lust for *Star Wars,* incidentally, we see that, oddly enough, the old print media is still a hidden and often unacknowledged engine for newer media, for it was the 1991 *Star Wars* novel, *Heir to the Empire,* which spent twenty-nine weeks on the *New York Times* best-seller list, that apparently triggered this revival "like a trip wire on the *zeitgeist*" (Handy 1997, 70). By 1997, there had emerged a totalizing campaign to further develop a market for *Star Wars* nostalgia. One "entertainment news" television program, *Access Hollywood,* covered the re-release by interviewing people just after they had seen the film. The interviewer's primary question was not to do with the film itself, but with the film in relation to the viewers' personal history (e.g., "I remember—I was about twelve years old when it came out the first time . . ."), thus making the film function as does the personal photograph or popular music—as a correlative narrative, a marker for personal time, and a tool for the fetishization of the past. In closing, the program's host, in nondiagetic narration, puts the phenomenon into perspective for us: *Star Wars* "brings us back to a time when good and evil were clearly defined." Of course, good and evil were probably no more definable in the Carter years than in the Clinton years (and incidentally, perhaps the reintroduction of a Democrat to the White House is related to the twenty-year nostalgia curve); it's simply that this is what is always said about the past, not only in the simplistic discourse of "entertainment news" but in academic discourse as well, if we will recall Anderson's notion of the "certitudes" that comforted the subjects of the now-vanquished agrarian age; and again, the ur-myth here is that of the Fall.

In any case, the strategy of the *Star Wars* revival—the strategy of exhuming a nostalgia-based film from the recent past—has become a standard strategy, as in the rc-release of the musical *Grease*, a film that nostalgically referenced 1960 from the vantage point of 1978, to be resurrected in 1998. Viewed in 1998, *Grease* can be experienced through the frame of "fifties nostalgia" or nostalgia for the young John Travolta, who reemerged in the 1994 film *Pulp Fiction* after a long career hiatus. This trend in entertainment, as indicated by this example, rests on the creation of multiple time frames for the audience, and this can become quite complex. To provide another example: George Lukas, who shrewdly held onto the licensing rights for *Star Wars* spin-offs, was engaged in the production of *American Graffiti* (1973) when he first conceptualized *Star Wars*. *American Graffiti* featured Ron Howard, who added to the film's nostalgic appeal by reminding viewers of his role as the fatherless child in *The Music Man* (1962) and "Opie" in *The Andy Griffith Show* (1960–68), both of which traded on the nostalgic myth of small-town America. In an action demonstrating the symbiosis of media entertainment systems, *American Graffiti* led to a spin-off television series, *Happy Days* (1974–84) (featuring Ron Howard as the central character, Richie Cunningham), and this series helped usher in the fifties nostalgia of the seventies. Because of rerun syndication, it was possible for a viewer in the late nineties to watch *Happy Days* with the following temporal frames: one, nostalgic longing for the 1950s; two, nostalgic longing for the 1970s (when the viewer first saw the program) and possibly a nostalgic longing for the 1950s nostalgia of the 1970s; three, nostalgia for "Opie." Further, As Lukas began work on *Star Wars*, he did so, as he said in an interview on the *Access Holly-wood* program, in view of his own nostalgic memories of the *Flash Gordon* series, which he personally wished to revisit. *Flash Gordon* (which, incidentally, is characterized to a large extent by the same past/future *mise-en-scène* as *Star Wars*), itself has the added twist of nostalgia for the vision of the future that was held in the past (as in the quaintness of the vacuum-tube technology as a pervasive sign of futurity in the *Flash Gordon* series).

It is in the peculiar generic forms that emerged in the later 1980s and the 1990s, in which the hands of time point both forwards and backwards, that we find the most advanced symptoms of what Carter calls America's "deeply rooted chronophobia" (1994, 36). The *Back to the Future* series, which comprises three films released between 1985 and 1990, is an example par excellence for a number of reasons. First, the series self-consciously locates itself within the heterochronic narrative tradition both through its oxymoronic title and through its visual references to the 1952 film version of Wells's *The Time Machine* (1895). In both films, timepieces are a central image (the clocks on the mantelpiece in the earlier film, the town's central clock in the latter) as is the outlandish machine itself (in the former, a kind of sled that resembles a piece of Victorian furniture; the later, the already comic Edsel of the 1980s, the Delorian).

Second, there is the way in which *Back to the Future* forms a homology with the economics of time. We noted earlier the importance of the shift from arbitrage to futures in the mid-nineteenth century—a shift from speculation based on space to one based on time. In his analysis of the relationship between *Back to the Future* and "Reaganomics," Nadel first points out that "credit is the theoretical mechanism by which time is turned into surplus value. To run out of time is to run out of credit, and to acquire credit is to gain time" (1997, 74). The Reagan economic solution (later denounced as a sham by Reagan's own budget director, David Stockman), which proposed massive tax cuts along with massive increased military spending, supposes the existence of an unlimited supply of credit, which was of course an unrealistic assumption. The entire premise of the *Back to the Future* series, which was one of the (if not *the*) most popular film series of the Reagan era, is the unlimited nature of the resource of time (credit), as exemplified, Nadel notes, in the scenes in which the Delorian achieves the conditions (at exactly eighty-eight miles per hour) to break the time barrier and in so doing expand time.

Third, and most importantly for our immediate purposes, there is the way in which *Back to the Future* renders one's personal relationship with time. Whereas in *The Time Machine*, the narrator/scientist is obsessed with probing the possibility of the moral evolution of the human race (in keeping with Wells's residual Victorianism), the main character of *Back to the Future*, Marty McFly, is concerned with time only to the extent that it relates to the domestic tranquility of his family and to some extent, his community (the post–World War II suburbs, given a Rockwellesque/Mayberryesque presentation).

The plot centers on Marty's return to the mid-1950s. He is faced with the task of straightening out the normal course of time, which has become disrupted, and he can do so by getting his mother and father (who, like Marty himself, are teenagers) to fall in love so that he can be born. Unfortunately, however, his mother abhors his father and falls in love with Marty himself. In one scene, she attempts to seduce the horrified Marty while parked in an automobile. Thinking back to our earlier discussion regarding the phenomenology of space and Bachelard's remarks of the security of enclosed spaces, we can see here clearly how modern nostalgia has replaced space with time, for Marty has returned in time to his own womb in the enclosure of the car, in dangerous proximity to his own birth canal, fulfilling the oedipal fantasy by temporally superseding his own father. Hoberman has referred to Marty as an "American Oedipus" who "works through his own family romance," the rewards for doing so being American enough: "an improved standard of living for his family" (1985, 48). The American "withness" in time and the entire preoccupation with nostalgia is evident throughout the series. Thus, in *Back to the Future II*, Marty, in his travels in the twenty-first century, visits a 1980s nostalgia "café," thus assuring the viewer in the present of the 1980s, when the film was made, that the nostalgia machine will keep on working into the future. This reminds us of Carter's examples of the wording of a NBC sports promo ("we're building tomorrow's memories today") and the assurances given by a radio DJ regarding a contemporary hit ("here's a future memory") (1994, 36).

A similar narrative-temporal strategy is employed in *Star Trek: Generations* (1994). In home-video form, the already nostalgic presentation is framed by another layer of commercial nostalgia. The promotional material that precedes the actual film includes a pitch for the "*Star Trek* universe of entertainment" through "unforgettable home video classics": seventy-nine classic episodes from the original *Star Trek* series (1966–69) featuring "Kirk, Spock, and all the characters families have admired for over twenty five years. . . . It's a nostalgic trip back to the future."

As the film itself begins, the retired Captain Kirk is treated like a media star as he attends the maiden voyage of the new starship *Enterprise*. The scene forms a odd exception to Kendall Walton's cardinal rules concerning the relationship of the "real" world to the fictional ones. According to Walton, "[C]ross[-]world" travel is impossible (17).

But this is precisely what happens here, for just as "Captain Kirk" is a legend according to the prescripts of the fictional world he inhabits ("I remember reading about your missions when I was in grade school," the new captain, Kirk's replacement, coos with admiration), so too is "Captain Kirk/William Shatner" a legend for the television viewer (and in fact, for the general American public). In other words, "Kirk" is a myth in the fictive world of *Generations* just as he is a myth in the "real" world of television viewing and the social configurations of fandom (*Star Trek* conventions, fan publications, etc.). Further, the obvious signs of late middle-agedness of the "Captain Kirk" of the fictional world is the selfsame agedness of "William Shatner" (which I put in quotes since the audience will know, not the historical William Shatner, but the television legend, "William Shatner"). The interaction between viewer and cinematext/videotext here erases the wall between world and fiction, for the viewer, if he or she had watched the original *Star Trek* when it first aired, will find that he or she has aged precisely as many years as have both "Captain Kirk" and "William Shatner." Feuer notes a similar mechanism at work in the musical comedy *Gigi* (1958), in which the aged actor Maurice Chevalier singing "I'm Glad I'm Not Young Anymore" expresses the sentiments of both the character and of "Maurice Chevalier" (Feuer 1993, 39).

This comparison of, on the one hand, the characters of our cinematic and televisual fiction as preserved in our memories with, on the other, the historical subjects who portray these characters seems to be a relatively stable element of the nostalgia market; thinking back to Husserl's notion of the that-which-is-not-present principle in perception, we might suggest that the same principle is at work here creating a particular aesthetic experience, that of the flashback-comparison, as a layer underlying the pleasures of the actual narratives. Consider, for example, Andrew Ferguson's playful observations on a nostalgically oriented television special; while gazing at an image of the cast of *The Brady Bunch* (1969–74) some twenty years after their currency as a popular television series and in tandem with the release of *The Brady Bunch Movie* (1995), Ferguson muses, "[T]hey were still recognizable, though barely. They seemed thicker, distended, as if someone popped a plug in the original kids and pumped them up to twice their normal size. The effect is grotesque" (1994, 80). The grotesqueness he refers to has little to do with the physical maturation of the historical subjects who comprise *The Brady Bunch*, for they have, after all, done nothing more than physically mature and age in the same manner as every other living being; the grotesqueness, rather, is a result of modern chronophobia, the state of being "outside of time" as Carter suggests, and the way that the technology of durative forms (e.g., videotape and film) creates frequent opportunity for such "grotesque" comparisons that highlight the disjunction between isolated and "captured" moments in time and the flow of time.

Following the introduction of "Kirk," the story moves seventy-eight years into the future to join the characters who were established in the second *Star Trek* television series (*Star Trek: The Next Generation*, 1987–94). We are presented with a group of officers preoccupied with a role-playing game of pirates walking the plank in the virtual reality of the holodeck, a heterotopia machine that allows one to enter the reality of one's choosing; thus, the characters engage in their taste for a nostalgic and romantic recasting of their own roles as navigators and rovers, while Picard mopes alone in his quarters, waxing sentimental and nostalgic over a set of family photographs (the fetishization of the photograph as a durative form vis-à-vis the vicissitudes of time is also a major motif in *Back to the Future*). And as if this is not enough, the plot that follows these establishing scenes is centered around a mad scientist type who wreaks havoc in the order of the universe in his effort to reinsert himself into a galactic energy ribbon, known as "the nexus," that maintains anyone within it in a state of timeless, nostalgic bliss, and where Captain Kirk now lives in a perpetual nostalgic past. Later, Picard too enters the nexus to deliver to Kirk the call to action (the first phase in Campbell's typology of the hero myth), whereupon the legend of Starfleet pulls himself out of his nostalgic stupor, returns to heroic battle in the real world, and meets a hero's death.

A similar chronoconstruction is found in *Apollo 13* (1995); the fact that a science fiction fantasy film and one that purports to more or less accurately follow a historical event in contemporary memory shows that chronophobia cuts across the entertainment spectrum. The film opens with a textual note placing us at the January 1967 Apollo I pretest accompanied by the disembodied but ever-recognizable voice of Walter Cronkite, whose authoritative-yet-avuncular ("Uncle Wally") persona, reminiscent, perhaps, of FDR, made his voice print something of a national aural symbol. "Inspired by the late President Kennedy," Cronkite intones, "in only seven years America has risen to the challenge of what he called the most hazardous and dangerous and greatest adventure on which our country could embark." The effect here is not unlike the opening line of *Star Wars*, "a long time ago, in a galaxy far away"—it is thus a mythopoetic rendering of Kennedy's "New Frontier" with all its Turneresque energy and with Cronkite as the Homer from what we're told was a golden age of television broadcasting before the news/infotainment era. Here we have an interesting twist: nostalgic longing (a romantic and sentimental feeling) for the purportedly more rationalist (i.e., nonromantic, nonsentimental) past.

The story then picks up on 20 July 1969, the date of Neil Armstrong's moon walk, and proceeds with the buildup to the Apollo 13 mission with a focus on the lives of the young astronauts (Jim Lovell in particular) via a montage of domestic scenes rendered with considerable attention to detail regarding the "look" of the period—furniture, clothing, hairstyle, etc. The popular music and images of the counterculture of the period are likewise referenced—snatches of Steppenwolf, Jefferson Airplane, Jimi Hendrix—but these signs of the period seem somehow disconnected from the narrative itself, as if they were only an ornamental backdrop. For the viewer, these ornaments provide an opportunity for nostalgia-based pleasure: the pleasure of observing those details of fashion that have fallen over the other side of a temporal horizon and are no longer part of "the contemporary." This approach falls under Graham's understanding of pseudohistory: "greased hair and high school 'mean' the 1950's. . . . the most historical is simply that which is the most 'historical'" (1984, 350). Angela Davis meditates upon this same problematic aesthetic in her essay on the reemergence of the Afro hairstyle as an element of fashion, totally "disconnected from the historical context in which it arose"—that is, the political struggles of black nationalists in the late 1960s. Davis makes particular comment on a fashion-advertising pastiche of her FBI-Wanted poster from 1970, in which her legal case is "emptied of all content so that it can serve as a commodified backdrop for advertising and thereby promote a seventies fashion nostalgia" (1994, 38, 41). In this neutering of what had once been some of the most politically threatening images to mainstream culture, we can detect both a playfulness aimed at an aesthetic pleasure and an example of what Milan Kundera, in *The Book of Laughter and Forgetting*, calls "organized forgetting."

A similar historical amnesia characterizes Howard's film, although here the matter is more ideologically problematic, since the send-up of Davis's Afro style, unlike *Apollo 13*, had no ambitions of presenting a legitimate historical narrative. In *Apollo 13*, the Vietnam War, which one would certainly expect to be a major part of any historical film dealing with the year 1969, is virtually absent. The entire focus of the film is on the teamwork between Mission Control and the Apollo crew in their joint efforts to get the damaged space capsule back to Earth. This is managed with an unrelenting reference to the team's winning attitude: unflagging optimism, technological ingenuity, a sense of "mission," a desire to cross frontiers, and fearlessness and humor in the face of death, all bolstered with a panoply of American clichés (e.g., "failure is not an option.").

Can we explain the ahistoricity of this allegedly quasi-documentary movie? Ron Howard's own acting career, as noted above, was heavily invested in the nostalgic narrative and what might be termed the nostalgia industry (indeed, when we think of the trajectory of Howard's career, we see that he enters the entertainment scene as a product of, and later becomes a producer in, the nostalgia industry), and as director of *Apollo 13* he chose the setting well. The year 1969 was proclaimed historic even before it was over. Journalists for the *New York Times* proclaimed the moon landing as the most significant event of the twentieth century; Norman Mailer called it "the climax of the greatest week since Christ was born," and Ayn Rand claimed that the most salient events of the summer of 1969 (the moon landing and the Woodstock Festival) were the perfect embodiment of Nietzsche's Apollonian and Dionysian impulses (Hoberman 1994, 10). Thus, the time frame, owing perhaps to the symmetrical symbolism of Woodstock /Apollo, had already been monumentalized before the moment had even passed. The film, then, takes this moment in time and the achievement of the space program and uses it mythopoetically. When we recall that Jean-Luc Nancy postulated that all history is conceived on the basis of a "lost community" that must be "regained and reconstituted" (qtd. in Bernasconi 1993, 3), the connection between Mayberry and Apollo—not to mention the epic tradition of Fenimore Cooper's Leatherstocking Tales, the *El Cid*, *Song of Roland*, Virgil's *Aeneid*—becomes clear. What is different in the case of *Apollo 13* is, again, the historical foreshortening of contemporary commercial nostalgia, for the mythic American past is conceived of as being a mere twenty-five years past. This makes history conceivable in terms of personal memory, and indeed the narrative is personally conceived, as the emotional trauma suffered by astronaut Lovell's wife and family is given a lion's share of screen time, while the Vietnam War has been virtually erased from history. But perhaps the most engaging and puzzling irony of the film is one that is central to American nostalgia and chronophobia: it presents us with a mythic past in which futurity is central.

Another salient feature of *Apollo 13* is the presence of the television; as mentioned, the film begins with the voice of Walter Cronkite, perhaps to incite the often-expressed nostalgia for a time when television news, at least purportedly, was seriously presented by credible individuals who projected a less glossy and jovial "personality" than the newscasters in the news-team format as formulated in the early 1970s as a response to increased network and local competition for viewers. Throughout *Apollo 13*, we are presented with scenes of people watching the events of that ill-fated mission as they were reported on television, and the response of the media to the space program forms a major subplot, making it seem as though the story is as much a remembrance of television as it is of a historical event—or rather, a remembrance of television qua mythohistory as the film works to promote reverie for a time when, purportedly, a news program could serve as the unifier of national consciousness and domestic togetherness, something that television and the media that preceded it always promised to be (as in the representations of the family huddled around the radio discussed in the previous chapter). Our knowledge that the film's director, Ron Howard, essentially grew up on television in the 1960s also may contribute a subtle phenomenal retention to the subject's reception of these scenes.

Relatedly, we see here that the media itself, rather than public fashion and historical events, has become one of the primary objects in the nostalgia market. The *Star Trek* movies, as noted, have been a major generator of this particular strand of marketing; *The Brady Bunch Movie*, as an entertainment magazine notes, is another example—a particularly illustrative one, since its central concept is to take a popular television series from the past and "drop the characters into the '90s!" (Jacobs 1995, 8). The film places the clean-cut, naïve, and sartorially backdated Bradys in a mid-1990s Los Angeles environment that is has purportedly degenerated into a morass of loud music, obnoxious and nihilistic teenagers, rapacious businessmen, and car-jackings, and the intended (or perhaps, attempted) comic effect comes from the Bradys' interactions with this new and threatening environment, as in their attempts to befriend a car-jacker. The misrepresentation of history here goes considerably further than in *Apollo 13*, for the film's "Brady-ness" has nothing to do with the mid-1970s: rather, the code of "innocence," which has been formulaically associated with the 1950s in contemporary commercial nostalgia, has simply been transplanted to the 1970s.

But the problem of referentiality here needs to be teased out just a bit further: it is less a matter of "Brady-ness" or "1970s-ness" that is at issue here; it is, rather, "television-ness." The 1950s generated two institutions that share a tight symbiosis: the nuclear family and the domestic comedy television genre. The new suburban family was to some extent produced by the economic and social conditions of post–World War II society. Extended family networks were rejected by the social work profession, committed as they were to "modern" ideas, and this position was aided by the housing boom, the development of suburbs (distancing families from relatives in the old ethnic urban neighborhood), and the standard single-family home. This new dispensation created enormous psychological pressures, for a relatively small unit, bereft of traditional resources, was now expected to produce "a whole world of satisfaction, amusement and inventiveness," an expectation that, in the history of the family, "had no precedents" (Coontz 1992, 27). (A sly comment on the impracticality of this family model is found in *The Brady Bunch Movie* in a number of sequences suggesting that Jan Brady has been driven to multiple personality syndrome as a result of the pressures of the nuclear family dynamic, specifically, intense sibling rivalry. This, incidentally, shows that popular entertainment can be in complicity and critical at the same time.)

Television, which had fully eclipsed radio as the major home entertainment genre by 1955, was soon in the business of providing idealized models of how these families were supposed to work in series like *Ozzie and Harriet* (1952–66) and *Leave It to Beaver* (1957–63). The feedback loop therein established is one in which families attempt, at least in part, to fulfill their mission of maintaining persistent entertainment and togetherness by watching programs about families. While it is hard to determine the extent of modeling that this loop encouraged or the degree of unhappiness that may have been caused, we can nevertheless, I think, agree with Coontz when she claims that "our most powerful visions of traditional families derive from images that are still delivered to our homes in countless re-runs of 1950s television sit-coms" (1992, 23). Arguably, these reruns, which enter television's "total flow" as a kind of filler for the margins that separate the more privileged time spaces of televisual programming, in the long run, through their particular phenomenology of reception, have proved more culturally central than the privileged prime-time programming. As Jenny Nelson explains,

> Reruns . . . spanning a period of forty years, mix freely in a sort of televisual twilight zone. As such, reruns form an amorphous cat-egory unique to the medium of television—one that encompasses and threatens the autonomy of specific genres and their specific time periods. The signifier "time period" is already beginning to lose a single, stable signified: does it refer to a programming slot, a period of program origination or of program reception, or a time period depicted in a program? And if it refers to all four, how does it do so? (1990, 81)

Nelson goes on to demonstrate that people tend to categorize televi-sion series according to personal time (e.g., programs watched during one's childhood), thus making television reruns a companion to popular music as an index of one's autobiography, the narratives of either song or program being to some extent irrelevant to the evocation of this nostalgic mood.

By the early 1980s, the Nickelodeon cable network capitalized on the unacknowledged importance of reruns by making them a high-lighted feature of its programming, sometimes including the original commercials that accompanied television series from the 1950s, 1960s, and 1970s. Rather than making specific marketing reference to senti-mental and personal associations with programs, Nickelodeon focused, and continues to focus, on a marketing strategy that stresses ironic pleasure in the outmoded behavior and fashion displayed in these pro-grams—though one wonders whether such irony merely conceals what might simply be sentimentality.

It is from this ironic turn that *The Brady Bunch Movie*, and other series in syndication recycled yet again into feature films, such as *The Flintstones* (1994), comes into being; the technology that made reruns possible, along with the other social and ideological factors that help to generate the nostalgia culture, constitutes a economic machine that produces recycled entertainments with what would seem to be unlimited energy. They suggest, as Allison Graham puts it, that "we are faced with the fact that our culture has become literally self-consuming, temporarily thriving on its own corpus" (1884, 354).

4

Visuality: The "Scopic Regime" of Popular Modernity

Having discussed space, voice, and time as they relate to experience under the phenomenal regime of popular modernity, we now turn to images and icons. This brings us back to some extent to our probing of spatiality, although here our emphasis will be upon visual entities (images and icons) contained in an overall spatial field rather than upon our sense of the field itself, and I will rely on somewhat different yet related analytical frames, particularly those theoretical areas that are situated between phenomenology and psychology, such as phenomenological psychology, semiotics (as practiced by, for instance, Kristeva and Silverman), and psychologies that deal with the self-other boundary.

The Image and the Boundary

The Renaissance discovery of the principles of visual-representational perspective is much more than a matter of art history; it is of central importance for the development of modernity in toto. Perspectivism and the realism of which it is a part inform an epistemology that defines the real as that which is visible and proceeds, like the Western hegemony itself, expansively and imperially, as both the microscope and the telescope proved eminently successful at disciplining the invisible to the rule of the visible. As Ivins states, the accomplishment of perspectivism is that it transformed vision from a mere form of "sensuous awareness" to the principal basis for all systemic, empirically conceived

thought: "[V]irtually every realm of technical innovation and manufacturing, from the onset of the machine age right through to the
information age, is predicated on the development of methods of pictorial understanding" (1973, 13). This transformation from the visual-
sensual to the visual-systematic is evidenced in the development of
mapping that we discussed earlier—in the transition from the medieval map, with its emphasis on the experience of place, to the post-
Renaissance map, with it's top-down, abstracted view of terrain.

The advent of photographic technology in the 1820s and 1830s
radically extends modernity's ocularcentric progression; Jay points to
1840 as the onset of a new "scopic regime of modernity," a new "reconfiguration of vision" (1988, 42). By the end of the 1840s, in a development that marks the transformation of modernity into *popular* modernity,
Americans were purchasing roughly three million daguerreotypes per
year, and by 1853, there were about one hundred commercial photography studios in New York City (Lubar 1993, 52). Throughout the
latter half of the nineteenth century in the industrialized West, photography was not only the defining image technology but also the diasporic
center for the entire range of Victorian spectatorship: the panorama
and diorama (of which Daguerre himself was master), circuses, magic
shows, industrial expositions, lithographic reproductions, and the ste-
reoscope (Slater 1995, 230; Richards 1991, 55–56).

And it is here we find that the "moment" of photography is not
only an establishing one for modernity but also one of reversal, for this
moment is one in which a positivist technical innovation (based on
rationalist science) becomes the foundational medium for a series of
representational practices that are based on illusion and fantasy. In short,
photography serves as the basis for a modern-cum-prerationalist popular
culture (cf. McLuhan 1962, 192–93). More broadly, this reversal—or
better yet, double movement—reveals the other side of the Enlightenment's project of demythification, a feature that we find repeatedly
identified: in Marx's claim that the industrial commodity is imbued
with the primeval religiosity of the fetish ([1867] 1978, 319–29),
McLuhan's notion of retribalization (1964, 262), Adorno and Horkheimer's argument that "the total effect of the culture industry is one
of anti-enlightenment" ([1944] 1993, 29), and Ellul's contention that
technological progress "restores man to the supernatural world from
which he had been severed" (1970, 191).

The relationship between ocularcentrism, technology, and the two

faces of modernity is best articulated by Don Slater in his study of photography as "natural magic," beginning with the camera obscura entertainments of the seventeenth century, which were not conducted as a
demonstration of the principles of physics and optics, but for their
supernatural effects. Slater defines "natural magic" as

> the power of science and technique at the height of their rationality
> to appear to us (who do not understand them) as a new form of
> magic. We believe in modern power over the material world, and
> even the most fantastic spectacles reinforce that belief: great spec-
> tacles demonstrate the power of science over appearances, and sym-
> bolize the power of science to transform the material world. The
> specifically modern nature of this spectacle lies in seeing the scientist
> as magician. (1995, 227)

In keeping with this, the photograph, in all its forms, from the family
portrait to pornography (both of which emerged almost immediately),
has an essential "magic," an ability to "capture time." As both Sontag
and Barthes point out, photography's uniqueness as a medium comes
from its ability to capture a "trace" of reality "directly stenciled off the
real, like a footprint or a death mask" (Sontag 1973, 154)—literally, a
photograph is "an emanation of the referent" (Barthes 1981, 76). This
"magic" connection to the real most likely fueled the photograph's
fetishistic value in both personal and public life, although an equally
"magical" quality is driven by its technological and scientific status,
evidenced by its "uniformity and repeatability" (McLuhan 1964, 175).

New techniques for making images, combined with mass production and the economics of industrialism, would come to have enormous consequences in the ensuing decades. Raymond Williams ([1980]
1993) suggests that in the generation that followed the economic depression that lasted from 1875 to the mid-1890s, it became evident to
the controllers of the economy that systemic stability required not
only the production and distribution of commodities but the introduction of an element of predictability in consumption. In order to
achieve this, workers had to supplement their productive role with a
consumptive one, and to do this, a fundamental cultural reencoding
was required—an ideological reorientation that allowed "excessive-
ness [to replace] thrift as a social value" (Ewen 1976, 25). Or as David
Harvey puts it, the shift from Taylorism (based on F. W. Taylor's

emphasis on labor efficiency) to Fordism (based on Henry Ford's five-dollar, eight-hour day introduced in 1914) was predicated upon Ford's "explicit recognition that mass production meant mass consumption, a new system of the reproduction of labour power, a new politics of labour control and management, a new aesthetics and psychology, in short, a new kind of rationalized, modernist, and populist democratic society" (1989, 125–26). In the United States, the Protestant ethic of thrift and the later development of a symbolic nexus for national political unity (as in, to use an example we're now familiar with, geopolitical maps used in compulsory education as well as advertising and general popular culture) are both reencoded into consumerist forms of duty and nationalism. This is precisely the process indicated by Debord when he aphoristically asserts that "the spectacle . . . is omnipresent—both as a reminder of one's duty to consume and as an instrument of unification" (1983, 3).

Such an ideological shift could only be mobilized psychodynamically, and hence the radical restructuring, made possible through applied psychology, of modern advertising, through which the primary spectacular power of "natural magic" was considerably enhanced. While the earlier mode of advertising, traceable to the mid-seventeenth century, is characteristically text-based, informational, and simplistic in terms of persuasive apparatus, modern advertising is informed by a "cultural pattern in which the objects are not enough but must be validated, if only in fantasy, by association with social and personal meanings." This is a "highly organized and professional system of magical inducements and satisfactions, functionally very similar to magical systems in simpler societies, but rather strangely coexistent with a highly developed scientific technology" (R. Williams [1980] 1993, 335).

Contrary to what is generally assumed, the cycle of consumption that capitalism strives to keep perpetually spinning is not propelled by materialism; it is not, that is, merely a shift from a rationalist "instrumental materialism" (the acquisition of goods for utilitarian motives) to a consumerist "terminal materialism" (the acquisition of goods as a autotelic motive), to use the terminology proffered by Csikszentmihalyi and Rochberg-Halton (1978, 7–8). Rather, the capitalist deformation of materialism is rooted in the abstraction that occurred in the split between use-value and exchange-value, a split that later, in consumer-based society, led to the "aesthetic abstraction of the commodity," in which the commodity's "surface appearance and its meaning detach themselves" and enter into an abstract relationship with "the character-masks worn by buyer and seller" (Haug 1986, 48–49). Debord concurs, claiming that the entire reformation of the individual vis-à-vis commodities under the conditions of late capitalism can be demystified (and here he echoes a prophetic remark by Feuerbach in 1843 that he cites at the beginning of his text) through the postulation of two phases, the first of which brings "into the definition of all human realization the obvious degradation of *being* into *having*"; this is followed by "the . . . total occupation of social life by the accumulated results of the economy" and the "sliding of *having* into *appearing*, from which all actual "having" must draw its immediate prestige and its ultimate function" ([1967] 1983, 17). Under such a system, we are not consumers of material commodities; "the real consumer," says Debord, "is a "consumer of illusions" (47).

Moving more broadly, then, to the question of experience and consciousness in the era of popular modernity: if, as Mitchell remarks, consciousness itself is to a large extent "an activity of pictorial production" (1987, 16), how does this activity take place in the specific context of consumer capitalism? We need to start at the cornerstone of capitalism—the commodity. The commodity, rendered as visual image and as represented in ideation, is, as discussed above, split from the utilitarian and materialist value of the object whereupon it forms relations with other image-concepts. One of these spheres is comprised of the subject's "self-image," also called "self-concept"; this terminological slippage in itself indicative of the overlapping boundaries of "image" (a specific visual perception) and "concept" (a nonvisual ideological construct), the relationship here being that of signifier and transcendental signified. The image:concept ratio allows ideological structures that cannot be accessed through primary experience to have a more tangible, sensuous form in the imagination, centering the self in a series of concentric circles indicated by phrases like "my image of America,"— formations that are supravisual and therefore more purely conceptual and ideological (cf. Boulding 1956, 3–14). And this, of course, brings us back to some of our earlier reflections on the admixture of ideology and the phenomenology of space in geography education. In any case, given these overlappings of visualization, self, and ideology (myths of the other, the nation, the corporation, etc.), the "Cartesian theater" of consciousness is perhaps best thought of, particularly in a spectatorially organized society, as a contested space in which images negotiate their interrelations and boundaries according to psychological and ideological constraints.

The concept of ego boundaries originates in the work of psychoanalyst Paul Ferden in the 1920s, though similar ideas are found in the work of Freud, Victor Tausk, and Josiah Royce, and later it would become an important element in Gestalt psychology, the Object Relations school of psychoanalysis, and the social sciences generally (cf. Gabbard and Lester 1995, 1–6, 12). For Ferden, the ego boundary is a function of an "ego feeling"—an ongoing sensation that supersedes the "boundary-less" state of infancy and provides a sense of a border that separates the "me" from the "not me," or as Ferden puts it, a sense of "how far the ego extends, or more correctly, the point beyond which the ego does not extend" (1952, 331); Anzieu would later refer to this phenomenon of the construction of body and selfhood as "the skin ego." In the accounts of Ferden and Anzieu, as well as those of Freud, Laplanche, Lacan, and others, the centered self is to some extent constructed from an internalized image of the surface of the body—the boundary between self and nonself (cf. Silverman 1996).

The structure of the self-concept has a curious relationship to any "objective," empirical seeing, such that even under ordinary circumstances it is questionable whether one ever sees oneself as others do; a pathological condition, anorexia nervosa, is a mere exaggeration of this "normal" condition. Furthermore, as we again reflect upon geopolitical maps (the map of the continental United States in particular), we find that the concept of the body politic is likewise conceived (and indeed, rendered conceivable) through visualized surface and boundary. It is also important to consider, with reference to our earlier discussion of space and habitat, the way in which the conceptualization of one's physical surface in conjunction with the conceptualization of outer conceptual layers—city, state, region, nation—reveals the interdependence of the conceptualization of ambient space and that of the visualizable entities that are contained by that space. Finally, this would mean that the image-orientation of the capitalist spectacle would come not from an "objective" capturing and rendering of surface, but from the ways in which visual conceptualization, particularly concerning the visualization and conceptualization of boundary, activates anxiety, fantasy and desire. During the formative period of modern commercial culture, a variety of spectacles, from the political to the promotional, were engineered to address the desiring self in terms of these contested boundaries and their attendant anxieties and fantasies.

Fascism, Consumerism, and the Scopic Regime

As one examines the iconography of the 1920s and 1930s, one is struck by the sheer quantity of counterphobic images—images that address the problems of defining, maintaining, and protecting the boundary of the self. I have thus far confined my remarks to the structure of consumer capitalism. There is also, however, considerable evidence of a connection between the iconic practices of the liberal consumer-democracy and the fascist state, for both systems were simultaneous projects in social and cultural engineering designed to solve the problem of predictability vis-à-vis social order in the industrial state, and as such, the images of both systems may be said to represent, if I may offer a modification of Jay's terminology, the scopic regime of *popular* modernity. Regardless of the overt political system, the totalizing organization of popular modernity in the industrial world is one in which "the formulators of the consumer market and the propagandists who publicized it hoped to instill an authoritarian obedience to the dictates of daily life in the machine age" (Ewen 1976, 96). The complicity between political systems with manifestly differing ideologies and what Ewen identifies as the underlying motive of modernity is further evidenced in the overall "design strategies" that are continually employed by both the state and the corporation. We find an instructive link in the very origins of modern organizational design; that is, the parallel development in Germany and the United States in the first quarter of the twentieth century of the interrelated notions of corporate image design and consumer engineering. In Germany, this move was advanced by the work of designer Peter Behrens, who provided an overall "look," products, logos, etc., for the corporate giant Allgemeine Elektricitäts-Gesellschaft (AEG); in the United States, these developments were advanced by designer Daniel Burnham and an early practitioner of corporate "image management," Earnest Elmo Calkins (Ewen 1988, 41–47). Interestingly enough, AT&T—a corporation that was primarily interested in the marketing of a form of communication devoid of both script and image—was the leader in the field of public relations advertising, creating and stimulating desire, and using image and text in these efforts (Fischer 1992, 72).

In the case of the fascist state, consider the iconology of Nazism. The rally at Nuremberg, September 1932, consisted of one million party members parading past the Führer with "32,000 flags and banners . . . the whole of the vast spectacle was enacted under a 'light dome' . . . formed by the vertical searchlight beams stabbing into a black sky" (Grunberger, qtd. in Lorentzen 1995, 163). Events such as these, along with the architecture of Albert Speer and the photography of Leni Riefenstahl (particularly the spectacle of human synchronization on a sublime scale, as in her series of photographs of German athletes performing calisthenics en masse in the Berlin Olympic stadium or the athletic scenes in her 1935 film, *Triumph of the Will*), are instances of an iconographic strategy, the purpose of which was to aestheticize the homogeneous "collective energy" of fascism (cf. Bataille 1979).

But the collectivist impulse in general, and nationalist and fascist impulses in particular, depends on the Other, the "not-me" that through contrast and opposition makes self-identity possible. In the case of pan-European anti-Semitism and the origins of fascist ideology, it was journalist Edouard Drumont (*France Under the Jews,* 1886, which was continually reprinted until 1914) who pronounced that anti-Semitism served as "the system of universal explanation . . . the negative pole of nationalist movements: it is in relation to the Jew, against the Jew, that the nationalist will define his . . . identity, proud as he is to belong to a community and to know clearly the adversary who threatens its unity and life" (qtd. in Winrock 1990, 137; cf. Eatwell 1996, 24). Mussolini's very late adaptation of biological anti-Semitism in 1938, after his earlier rejection of biological theory and his ambivalence regarding making a break with Jewish-Italians and going against the antiracialism of the Roman Catholics (cf. Michaelis 1978, 163, 187), indicates his understanding of fascism's reliance on the Other, not because of what the Other necessarily is, but because of how it can generate collective energy; as he said, "the fight against these powerful forces [the Jews], *as many consider them to be,* serves to give the Italians a backbone" (qtd. in Bernardini 1982, 30).

We see, then, that one of the ways that fascism performs its collectivist function is through the construction of both boundaries and what is beyond them, and indeed, using the apprehension of the Other to further consolidate the collective. Thus, to return to the German example, the aesthetics of Nazism performs a counterphobic function in terms of the fear of the invasion of self—the invasion of the German body—by the petulance of foreign bodies, both cultural and physical. The complicity of this aesthetic principle with the imperatives of the state is evidenced in the official Nazi policy (initiated in 1934) against the more destabilizing aesthetics of radical modernist practice (cubism, expressionism, primitivism, and other forms of "degenerate art," all of which had considerable traffic with the more radical and antibourgeois fascism of the Italian futurists). The Nazi retreat into the "orderliness" of a debased classicism (with its clearly delineated and racially "pure" bodies, as in the sculpture of Arno Breker) is but another manifestation of the fascist specular totality.

In *Male Fantasies, Vol. I.*, a study of the writings (diaries, novels, letters, etc.) of pre-Nazi Freikorpsmen, Klaus Theweleit notes the tendency to refer to the Bolshevik (and Jewish) threat in terms of fluids— "the wave" of Bolshevism, the "red flood," the "raging Polish torrent" (1987, 229), and he implies that these "threats" to the male ego may be symbolically controlled through the use of masculinist icons ("'erections': towering cities, mountains, troops, stalwart men, weapons" [402]), which in iconic form define the male self. These images of order form one essential axis of popular modern iconography, although there are deep, premodern roots. In the very origins of the state, Paglia notes, order and the *imaging* of order emerge simultaneously; in the "grandiose and self-divinizing" icons of the Egyptians, "social order becomes a visible aesthetic, countering nature's chthonian invisibilities" (1990, 185). Debord would concur, for he states that the oldest social specialization is that of power, and that power is at the very core of the spectacle. It is "the existing order's uninterrupted discourse about itself, its laudatory monologue. It is the self-portrait of power . . ." (1983, 23, 25). This was well recognized by seminal fascist ideologues such as Georges Sorel, who, in his *Reflections on Violence* (1908), opposed rationalist political theory by positing a politics of myth and claiming that national myths were indivisible image systems that had to be grasped by the masses intuitively. Another French ideologue, Robert Brasillach (who was active in the 1930s), was one of the first to note that the incipient success of Nazism was rooted in its powerfully suggestive images, which amounted to "a kind of poetry"

(Sternhell 1995, 78, 250). George Mosse, commenting specifically on twentieth-century fascism and its aesthetics, thinks similarly: "[I]n fascism, power had to express itself visually" (1996, 245). As evidenced by the history of iconographic practices, from ancient totalitarian regimes to the Counter Reformation, the power of the visual image was both recognized and exploited long before the emergence of rationalism, and indeed the image strategy of consumer capitalism is but a post-Enlightenment continuation of such usages and an evidence of what Herf, in his study of Nazi ideology, calls "reactionary modernism."

In the case of the liberal consumer-democracy the historic/iconographic rupture occurs in a somewhat different manner. It is significant that during the formative period of modern advertising, a broad variety of products capitalized on the fear of bodily contamination and social rejection due to disorders like halitosis, "acid mouth," "brain fatigue," acne, and so on. Products like Listerine mouthwash, Lifebuoy soap, Ivory soap, Lysol disinfectant (which warned in a 1926 ad that "innocent looking objects may be fraught with the dangers of unnecessary illness") were particularly dependent on this kind of phobic advertising (cf. Edgar Jones 1979, 224, 225, 260, 282, 297). Meanwhile, the telephone industry (both AT&T and independents such as the Illinois Telephone Association) used the fear of criminal invasion of the home to sell the telephone as a security technology, allowing one to contact the outside world for help (thus we see that phenomenal orientations to media can be directed in various ways, for earlier we discussed examples of how the telephone was itself the invader) (cf. Fischer 1992, photo 10). And while much of this was no doubt driven by legitimate public health and safety issues, it was also a response to the economic need to establish new consumer markets by identifying latent fears and providing a commodity solution, increasingly through the use of associative imagery rather than through rhetoric and text. And as in the German example, there is a metonymic relationship between the individual physical body and the body of the state regarding these anxieties, as revealed by parallel obsessions to promote the "American Way of Life" and to maintain the appearances demanded by this vaguely defined ideology, to be a good citizen in what Roland Marchand calls the "Democracy of Goods" (1985, 218). All of these obsessions are perhaps best illustrated by the now infamous and amusing advertisement for Scott paper towels, "Is your washroom breeding Bolsheviks?," in which the fear of foreign contamination and the fear of bacterial contamination are revealed as having a common source (fig. 4).

The Scott Paper ad also points to another item of significance, for the spectacle confronted by the disgruntled employee in the ad is one of disorder—indecent, "awkward . . . unsanitary" conditions. In modernity's specular totality then, we shall find not only images of order but also an equivalent iconic focus on images of disorder; again, as in the fascist aesthetic, it is the alien Other (in the ad, the specter of Bolshevism) that makes the self and the collective meaningful. Indeed, in Theweleit's study, the images of destruction and disorder are inexorably wedded to those of soldierly order, for in Freikorps literature the alluring yet destructive "red" women (Jews, Bolsheviks, Poles, etc.) are described as "animals" who stir the emotions (i.e., violate the boundaries) of the soldier-males and therefore must be violently eradicated, "cut to pieces," "drenched with black blood," reduced to "bloody masses" (1987, 183–204).

Another important element of Freikorps literature is found in its emphasis on homosocial camaraderie. In, for instance, Ernst Jünger's *Storm of Steel*, a novel depicting the author's experiences as a soldier in the kaiser's army during the First World War, the importance of homosocial bonding is evident. In the novel's conclusion, Jünger muses on the importance of these bonds:

> Hardened as scarcely another generation ever was in fire and flame, we could go into life as though from the anvil; into friendship, love, politics, professions, into all that destiny had in store. . . . We stood with our feet in mud and blood, yet our faces were turned to things of exalted worth. . . . It was our luck to live in the invisible rays of a feeling that filled the heart, and of this inestimable treasure we can never be deprived. (1929, 282–83)

In this purple passage, masculinity is something ethereal and ideal— it "rises above" the feminine earth with its mud and blood (its excrement and menses), and the entire verbal image, like the image in the Scott Paper ad, is based on a binary structure of order:disorder organized around a lone male figure (the plant manager, symbol of corporate order; the soldier, the symbol of state order) who confronts this boundary. The masculine body in Jünger's passage is furthermore an armored body (metal tempered in "fire and flame") and cased within

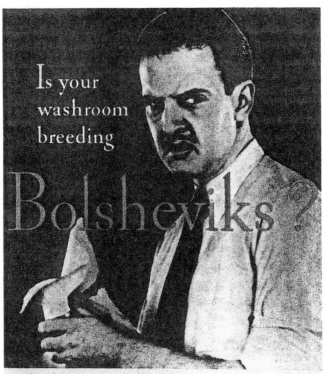

Is your washroom breeding Bolsheviks?

Employees lose respect for a company that fails to provide decent facilities for their comfort

TRY wiping your hands six days a week on harsh, cheap paper towels or awkward, unsanitary roller towels—and maybe you, too, would grumble.

Towel service is just one of those small, but important courtesies—such as proper air and lighting—that help build up the goodwill of your employees.

That's why you'll find clothlike Scot-Tissue Towels in the washrooms of large, well-run organizations such as R.C.A. Victor Co., Inc., National Lead Co. and Campbell Soup Co.

ScotTissue Towels are made of "thirsty fibre"... an amazing cellulose product that drinks up moisture 12 times as fast as ordinary paper towels. They feel soft and pliant as a linen towel. Yet they're so strong and tough in texture they won't crumble or go to pieces... even when they're wet.

And they cost less, too—because one is enough to dry the hands—instead of three or four.

Write for free trial carton. Scott Paper Company, Chester, Pennsylvania.

ScotTissue Towels - *really dry!*

Fig. 4

Washroom Bolsheviks. Advertisement for paper towels (ca. 1936).

that armored body is another—hermetically sealed blood, uncontaminated—in which the ideal mysteriously exists. We see in this passage the general ideological character of fascism: it rejects both materialism and rationalism in the name of a quasi-mystical and mythopoetic idealism.

The various texts cited by Theweleit constitute a literature that forms an odd kind of cognate to the horror literature that developed in the liberal democracies, particularly in the United States, at about the same time. The work of H. P. Lovecraft, whose major period of production was in the 1920s and who is considered a seminal influence on modern horror, provides the most effective example. A typical Lovecraft tale, "The Lurking Fear" (1925), illustrates as well, if not better than Theweleit's accounts, the soldier-male ideology and its psychosexual underpinnings, particularly in the relationship between gynophobia, homosociality, homosexuality, and homophobia. The plot is concerned with the attempt of the narrator and his cohorts to find the secret behind a series of grotesque murders among the poor folk in an obscure region of New York State. As in the Freikorps literature, there is considerable homosocial camaraderie between the narrator and his Theweleitian "soldier males," or as Lovecraft's narrator puts it, "faithful and muscular men" (Lovecraft 1971b, 1). The imagery of one of the first scenes suggests a military mission as the narrator and his comrades enter the region, barricade themselves, and survey the landscape:

> [T]he three of us dragged from another room a wide four-poster bedstead, crowding it laterally against the window. Having strewn it with fir boughs, all now rested on it with drawn automatics, two relaxing while the third watched. From whatever direction the demon might come, our potential escape was provided. (5)

The scene places these three soldier-males in bed together, symbolically erect ("automatics drawn"), and the potential intimacy of is heightened when the narrator, after describing his place on the bed between his two companions, tells us that he was awakened when his companion "restlessly flung an arm across my chest" (5). We are reminded that Theweleit and many before him (e.g., Freud, Reich, Adorno) have commented on the homosociality of the soldier-male, as in Adorno's axiomatic claim that "totalitarianism and homosexuality go together" (qtd. in Theweleit 1987, 55). Although Adorno takes the point too far, conflating homosexuality and homosociality (and perhaps

driven by homophobia), we can postulate that in the fascist male imagination there is both a loathing for women and a loathing for men, for
to travel too far down the path of homosociality raises the specter of
same-sex desire, which is unacceptable to the fascist because it poses
the threat of allowing oneself to be feminized. In horror and in fascism, both women and homosexuals are monstrous (cf. Grant 1996;
Benshoff 1997). In the above scene, the slippage between homosociality
and homosexuality is evident; and following this, the slippage between
homosexual imaginings and homophobia likewise becomes manifest.
We should observe that the narrator is awakened not only by the arm
over his chest but by disturbing dreams, thus equating male:male contact with psychic disturbance.

Further, immediately after the companion had flung his arm across
the narrator's chest, both of the narrator's companions are destroyed
by the "lurking fear," at this point only identified by it's "hideous . . .
shrieks" (the shrieking harpy of the male gynophobic imagination?).
Later, the narrator acquires yet another male companion, and again,
shortly after the narrator engages in male-to-male contact—he touches
his companion on the shoulder and "playfully [shakes] him"— it is
discovered that this companion too, has been destroyed by the lurking
fear (Lovecraft 1971b, 11). In these descriptions, then, we can find a
combination of male and female sexual imagery in the context of psychic discomfort, primordial disgust, and punishment.

We have already mentioned Theweleit's observation of the metaphoric concatenation of Bolshevism and Jewishness in terms of "floods"
and "torrents." In various places in Theweleit's source texts, the "red
flood" of Bolshevism is described as a "swamp" or a "morass." Turning to our story, as the narrator continues his quest for "the lurking
fear," the landscapes are feminine in terms of imagery, and biological
fecundity is repeatedly described as a source of loathing and nausea.
The surrounding forest is "morbidly over-nourished" (Lovecraft 1971b,
8); the abandoned gardens are "polluted by a white fungus, foetid,
over-nourished vegetation that never saw full daylight" (12). These
landscapes create a sense of dread and loathing in the narrator: "I could
trace the sinister outlines of some of those low mounds" (12); ". . . I
hated it. I hated the mocking moon. . . the festering mountain, and
those sinister mounds. Everything seemed to me tainted with a loath-
some contagion . . ." (18). Thus, there is a connection between the
feeling of dread and loathing felt by Theweleit's Freikorpsmen and

Lovecraft's narrator as he gazes upon the fecund earth. The narrator's efforts to find the beast (which turns out to be a de-evolved human, a "filthy whitish gorilla thing" [22]) leads him to the moldering earth underneath the fecundity on the surface; he forces his way into an underground burrow (as with Jünger's heroes who must stand in muck and mud); he becomes obsessed by desire, by "a mad craving to plunge into the very earth of the accursed region" (17) (here perhaps more an image of the male anus than of the female vagina), with regard for neither "reason [nor] cleanliness" (16) until he finds "the lurking fear":

> [T]he thing came abruptly . . . a demon. . . . Seething, stewing, surging, bubbling like serpent's slime it rolled out of that yawning hole, spreading like a septic contagion. . . . Shrieking, slithering, torrential shadows of red viscous madness. . . ." (20–21)

And thus again, the imagery of this story reveal a polymorphously phobic condition regarding sexual contact.

But there is also a political subtext to "The Lurking Fear." The creature that is the object of their quest is a racial degenerate, and even the peasant population are described in such terms: "[S]imple animals they were, gently descending the evolutionary scale because of their unfortunate ancestry" (8). The narrator is referring to the local landed aristocrats, descendants of a seventeenth-century New Amsterdam settler who had "disliked the changing order under British rule" and did all he could do to isolate himself and his kinfolk from it to such an extent that all his descendants up to the present (the story is set in 1921) "were reared in hatred of the English civilization" (13). Given this, the struggle between the narrator and his aides and the mysterious "lurking fear" expresses both sexual anxiety and political/territorial conflict.

Lovecraft's overt politics have much to do with the subtext of "The Lurking Fear" and other tales from the Lovecraft corpus. Lovecraft's racism, xenophobia, and admiration for Hitler are amply documented (cf. De Camp 1975, 374), and critics have pointed to an informing fear of "racial miscegenation [and] inbreeding" along with a "roiling xenophobia" in a number of his tales, particularly in "The Shadow Over Innsmouth" (1936) (Lovett-Graff 1997, 175). On one occasion, Lovecraft referred to the immigrant inhabitants of the Lower East Side of New York as "swine . . . a bastard mess of stewing mongrel flesh without intellect, repellent to the eye, nose and imagination—

would to heaven a kindly gust of cyanogen could asphyxiate the whole gigantic abortion, end the misery, and clean out the place" (qtd. in De Camp 1975, 168). Such a statement makes the subtextual elements of "The Lurking Fear" unmistakable, for the description of the biological degenerates of Lovecraft's fiction is starkly similar to his commentary on the American immigrant population.

There is far more at work here, however, than H. P. Lovecraft's overt political and racial ideologies, for the narrative structure of "The Lurking Fear" is a typical one in the horror genre. An earlier and very significant example of this structure can be found in one of the primary texts of the horror tradition, Stoker's *Dracula* (1899), which Sali Klein has referred to as a "pre-fascist fantasy" (1992, 275). In the plot, which is very similar to that of "The Lurking Fear," Dr. Van Helsing and his band of doughty men endeavor to destroy the unnatural creature, a foreign invader from the mysterious East who threatens to spread its progeny throughout England, thereby constituting a threat to the purity of English blood (cf. Halberstam). Indeed, the body of Lucy Westenra ("Westerner," as Twitchell points out [1985, 129]), Dracula's first victim, is a site where these anxieties find expression. In an early part of the story all of Van Helsing's soldier-males contribute their blood to Lucy's body, which has been contaminated by the vampire. But the procedure fails; she has been contaminated beyond redemption, and as they gaze on her now-changed features, they see that her "purity" has been changed to "heartless cruelty" and "voluptuous wantonness" (Stoker 1978, 217). This unnatural woman, the soldier-males discover, has been thoroughly polluted by the foreign invader, and ultimately they must destroy her. First, they impale her with a wooden stake, with her betrothed, Arthur Holmwood, doing the honors. Stoker's prose here is remarkably Jüngeresque:

> [Arthur] looked like a figure of Thor as his untrembling arm rose and fell, driving deeper and deeper the mercy-bearing stake, whilst the blood drom the pierced heart welled up around it. His face was set, and high duty seemed to shine through it; the sight of it gave us courage so that our voices seemed to ring through the little vault. (222)

But this is not enough. Not only must they impale her, but symbolically castrate her as well (Van Helsing insists that Lucy must be decapitated as a final solution).

Horror literature is largely concerned with biological boundaries and the sense of revulsion that accompanies our encounter of the disorder of the corpse. As Noël Carroll has observed, the threat posed by the creatures of horror is generally "compounded with revulsion, nausea, and disgust. . . . [They are] putrid or moldering things, or they hail from oozing places . . ." (1990, 22–23). The slippage, then, between the texts of horror and other texts is based on the extension of this primal disgust to the symbolic boundaries of the self as reflected in the ideologies of nation, race, and gender.

"The Lurking Fear" and other related horror narratives, then, is structurally homologous to the Freikorps and Nazi battle against the other—female, Polish, Jewish, Bolshevik, homosexual—those things that the fascist imagination imagines as threats to boundary and order. The fascist imagination and the imagination of horror share this fear of the other, and the heroes of both narratives succeed insofar as they steel themselves against this spreading decay and ultimately destroy it.

The Margins that Form the Center:
Fascism, Horror, Pornography

John Berger states that "in no other form of society in history has there been such a concentration of images, such a density of visual messages" (1972, 129). The society of which he speaks is that of the industrialized nations of the West after the Second World War—after the virtual elimination of overtly fascist regimes from the First World and the establishment of a "containment" policy for the alternatives to capitalism. During this period American-style consumer culture came to the forefront not only in the United States, in which a strong economy and a newly generated youth market fueled the "image industries," but in Europe as well. This process began in the first decade of the twentieth century, when the United States began to establish an economic-entertainment hegemony in Europe, best demonstrated in the struggle for dominance in the emergent motion picture industry. One prophetic European observer noted in 1910 that "the Americans will soon conquer the European market and impose their regulations on Europe" (qtd. in Puttnam 1998, 41; cf. Schiller 1969, 1–17, 79–87). In the years following World War II, the American occupation become the launchpad for the hegemony of American popular culture, particularly popular music. By 1964 Great Britain had become a formidable competitor to the United States, as it become obvious that British economic prosperity might well depend on commercial markets in the United States, Canada, Australia, and other parts of the Anglophone and non-Anglophone world. And thus the 1960s witnessed a British "invasion" of North American entertainment markets, which manifested itself in rock bands like The Beatles and The Rolling Stones as well as in the James Bond phenomenon, which well illustrates the British shift from military to fantasy hegemony, as Bond keeps the world safe for the British Empire while living according to the new and decidedly Hefneresque values of "swinging Britain," with its Americanized values of "classlessness and modernity" (Bennett 1983, 202). These developments were abetted by the deployment of more liberal consumer credit policies, more sophisticated approaches to advertisement, and the further saturation of public life in mediated images, largely through television. These developments were later followed, in the 1980s, by the triumph of the consumerist state in Eastern Europe, the Soviet Union, and, to a greater or lesser extent, other parts of the world. Media saturation began in the preindustrial world and, concomitantly, Japan arose as an important supplier for the technologically driven entertainment industry. And in these contemporary image industries there is a singular persistence of the concerns for the boundary of Self and Other announced in the iconographies of fascist art and Lovecraftian horror.

Lovecraft's work is absolutely seminal to late-twentieth-century novelistic and cinemagraphic horror, to the work of Clive Barker, Stephen King, and Wes Craven, to name a few. It is also important to note that attempts to establish a cordon sanitaire between Lovecraft's ideology and the culture of horror should arouse our suspicion: while one need not be a fascist to write horror stories, and while one should beware of overgeneralizing about this genre (cf. N. Carroll 1990, 197), one must nevertheless have an understanding of the way this genre is propelled by the soldier-male and by the fear of the Other, in particular that which is rendered as feminine.

A popular American film released in 1996—*Independence Day*—demonstrates the persistence of the horror spectacle and its fascistic subtext. This film—polygenerically constructed from the conventions of science fiction, nationalist war narrative, domestic melodrama, and action-adventure—is yet another retelling of the fantasy of a hostile invasion of the planet and how that invasion eventually fails. First threatened by this invasion are the symbols of the American Empire—one of the film's first images is that of the massive alien "mother ship" casting its shadow over the American flag planted on the moon by Neil Armstrong in 1969, thus indicating that this threat has crossed the outermost boundary of the empire (and also indicating the continuing imagistic and mythic power of the Apollo 13 mission, as discussed earlier). Soon thereafter, we are presented with a visual catalog of American architectural icons—the Iwo Jima statue, the Lincoln Memorial—and thinking back to Hobsbawm's remarks regarding the invention of the modern nation-state, we'll recall that national monuments (along with celebrations and public education) form the tripod of nationalist sentiment and ideology (this notion is given a complex extra layer here, for as Rogin notes, as an election-year movie that received the support of both candidates, Clinton and Dole, *Independence Day* itself became something of a nationalist monument [1998, 9]). The second entity that is the object of threat is the nuclear family, as the imagery of American nationalist monuments is compounded by sentimental domestic scenes in which the major male characters are portrayed as trying (successfully or unsuccessfully) to be good husbands and/or fathers (as noted in chapter 1, there are two primary American spatial myths—one concerning national spatiality, one concerned with the domicile). In keeping with the multicultural imperatives of recent popular culture, this attempt at a cross-section of American life consists of a number of social classes and ethnic and postethnic types: white politicians and bureaucrats based in Washington, D.C., New York Jews, upwardly mobile blacks, poor whites, one Hispanic, and one gay male. This multicultural smorgasbord is then given a global rendering as we are presented with alien sightings from around the world done up as a *National Geographic*-style panorama of "colorful" Others as viewed from the Middle Earth of North America.

The mother ship then sends a squadron of enormous flying saucers into the Earth's atmosphere, each one occupying the airspace over a major city. In a series of significant images, a central disk on the underside of the saucers aligns itself with major features of the aforementioned monuments; that is, the central disk of the saucer over New York City hovers directly above the Empire State Building (certainly a kind of monument to capitalist energy); in Washington, D.C., over the Capitol Building. When the aliens begin their attack, the central disc opens up to reveal a central orifice that gives forth a numinous, sublime, bluish glow—"It's beautiful," exclaims one of the witnesses just as the orifice reveals itself as the portal for a destructive weapon. It fires a ray that makes the monuments implode, thus signaling the onset of the invasion.

While it may seem premature, or an instance of psychoanalytic overkill, to proclaim the "mother ship" and these mysteriously glowing orifices that open up to destroy the grandest phallic projections of American architecture as instances of the archetypal destructive mother and of the fear of the feminine that we have discussed vis-à-vis Lovecraft's fiction, such an interpretation gains warrant in the film's most horrific images—indeed, Lovecraftian images—which occur in the scene in which government scientists at the quasi-mythical Area 51 in New Mexico probe the body of a captured alien, who, like the creatures from the *Alien* series (1979–97) and other contemporary sci-fi horror films, is designed to engender primal disgust. The chief scientist, as though performing a cesarian section, slices open the "biomechanical exoskeleton" of the alien's skull, which then pops open to reveal a gelatinous mass of cellular matter that exudes a powerful and disgusting smell. He reaches in, parting the folds of gelatinous flesh, and finally, a flap of slimy tissue that covers the alien's actual skin.

Following this, the alien awakes and begins his attack on the humans. When the president, in the best of liberal democratic traditions, attempts to negotiate with the creature, asking him "What do you expect us to do?," the alien, whose appearance is now that of a small black devil with glowing eyes, replies: "Die." We are then told that the aliens are purely destructive; like locusts, they move through the galaxy, destroying civilizations and recklessly consuming resources without regard for other life forms (which actually sounds pretty close to American civilization; it brings to mind the maxim that our enemies are our enemies precisely because they are so much like us, which problematizes considerably the Self:Other dynamic).

Beneath *Independence Day*'s purported political correctness and multiculturalism, in which all Americans struggle valiantly with the problems of relationships and domesticity (but which, in fact, end up supporting male dominance in these relationships [Rogin 1998, 44]) and in which all the people of the world ultimately unite (under the leadership, of course, of a white North American), we find that there must always be the excluded Other, the creature that is so evil that there is nothing redeemable about him and who must be destroyed in violent and heroic battle. The characterization of the aliens as a kind of pure evil links this particular entertainment with a long tradition of stigmatizing the other, particularly the Jew, and the parallel works here in a number of ways: references to callousness, migratory and predatory behavior, and physiological repulsiveness.

The notion of the Jew as a manifestation of callousness can be found in the words of Luther: "[A] Jewish heart [is] as hard as wood, as stone, as iron, as the Devil himself. In short, they are the children of the Devil" (qtd. in Morais 1976, 153). Similarly, the alien in *Independence Day* is presented, as we noted, as a small black devil who is relentlessly aggressive and has no pity for the human race. He only wants them "to die."

The president, after experiencing a mind-meld with the aliens, has an insight into their nature as migrants continually moving through the universe—we will recall here the notion of the Jew as the ultimate alien, a nomad whose homelessness is part of a punishment for his or her evil nature, as in the medieval myth of the wandering Jew with its origins in a particular reading of Deuteronomy ("And the Lord will scatter you among all peoples" [Deut. 28:64–66]). (But as this is an American story, we might well also return to our ruminations on Turner: the aliens are the marauding redskins, who, for all their technological superiority, stand on the other side of that frontier which separates "civilization" from "savagery.")

In his study of the psychodynamics of anti-Semitism, Mortimer Oston catalogs the crudest elements of this time-honored bigotry, which holds that Jews are "ugly . . . greedy bloodsuckers . . . [who] smell [bad]" (1996, 131). This was an obsessional idea with Hitler, who complained in *Mein Kampf* (1925–26) that the smell of Jews made him sick to his stomach. In *Independence Day*, there is continual reference to the alien's obnoxious odor, and we will recall similar statements made by Lovecraft about Jews and other urban immigrants, as well as the sentiments of white supremacists, who claim that blacks have an offensive odor. The presence of these obsessional delusions regarding disgusting odors is rooted, like the horror genre itself, in the fear of the corpse, which is then transferred, as noted earlier, to other entities beyond the boundaries of the self, including the national self as rendered in the theory of scientific racism. It is also related to an animalistic mythos that suggests that we may be able to sniff out danger.

The human race's counteroffensive is initiated as the president delivers a stirring speech to fighter pilots, telling them that the Fourth of July will no longer be an American holiday, but rather a world holiday. This movement in the film from a nationalist ideology to a momentarily and questionable internationalist one also has an uncanny parallel in the history of anti-Semitism. In a number of his speeches to the Reichstag, Hitler referred to Jews as "parasites living on the body of the productive work of other nations" (qtd. in Bayner 1969, 740). He advised that

> If the international Jewish financiers in and outside of Europe should succeed in plunging the nations once more into world war, then the result will not be the bolshevization of the earth, and the victory of Jewry, but of the annihilation of the Jewish race in Europe. (740)

And elsewhere:

> Only when the Jewish bacillus infecting the lives of people has been removed can one hope to establish a co-operation amongst the nations which shall be built up on a lasting understanding. (741)

It is furthermore noteworthy (returning to the psychosexual axis of this interpretation) that the most downtrodden of *Independence Day*'s American male protagonists—a working-class, down-on-his-luck, divorced or abandoned, alcoholic Vietnam veteran who claims to have been sexually abused by aliens—is ultimately the sacrificial hero of the story. In the turning point in the war with the aliens, the intrepid veteran, promising to deliver "payback" to the aliens (for having been sodomized?), pilots his fighter jet into the alien craft's destructive orifice while shouting "assholes" and "up yours." As Rogin puts it, the film's "climax . . . allies anal penetration with death, escapes maternal power and rescues heterosexuality" (1998, 71), fulfilling the psychosexual desiderata of the soldier-male. The Vietnam veteran, a stock figure representing feelings of failure and demasculinization, has thus gained military victory, regained the respect of his son, and remasculinized himself as the phallus that destroys the destructive cunt of the threatening mother.

Moments after the veteran's death, his son Miguel is told his father was a hero, and he should be proud. And of course he is, and thus the racial Other is assimilated by the national body, first biologically (he is apparently the son of an Anglo father and a Mexican mother—yet another female Other that must be overcome) and second, ideologically. Meanwhile, two other soldier-males (a Jewish computer genius and a black action hero, stereotypically rendered and in the service of the white president) use a captured alien craft to surreptitiously fly into the mother ship coordinating the invasion from outer space; they pass through its inner organic-looking structures, deliver a nuclear payload, and exit before its entry orifice has time to close them in.

In its homosocial nationalism, its internationalism predicated on the demonization of the Other, and in its fear of the feminine, *Independence Day* stands as a prime example of the continuation of the Lovecraftian tradition of fascistic iconology; more importantly, this film's reiteration of the horror/fascist narrative is by no means unique. The soldier-males of Stoker and Lovecraft and of *Independence Day* are evident throughout the development in the 1970s and 1980s of action-adventure heroes whose body imaging projects a particular kind of masculinist idealization. This body ideal is very closely connected with that relatively new icon, the bodybuilder (who often becomes the action-adventure actor or professional wrestler, something of the same thing, as a midcareer switch). Bodybuilding, throughout most of its brief history, was a marginalized, even deviant, iconic practice, but by the 1980s it became part of mainstream male iconography, and hence the action-adventure hero could, for the most part, no longer be the paunchy John Wayne or the somewhat slender Roger Moore, for masculinity and heroism would increasingly be defined by body image. Of the forms of this particular tradition of the national hero, traceable to the *Aeneid*, *RoboCop* (1987) is perhaps the most effective; it also contains modern fantasies of the metal body. We should be reminded here of Jünger's phraseology: true men are "hardened . . . in fire and flame," able to "go into life as though from the anvil." The initial *RoboCop* film spun a vast commercial web that ultimately included not only two sequel movies, but also a comic book series, an arcade game, a "Gameboy" program, a Halloween costume, toys, a watch, a sheet and pillowcase set, and children's underwear (!), all bearing the image of RoboCop with his expressionless, eyeless gaze and his bulky metallic exoskeleton (even the portion of his flesh that we can see, his jaw, is hard—a "granite" jaw).

RoboCop, The Terminator, Judge Dredd, the "Mighty Morphin Power Rangers" (ordinary teenagers who can collectively transform into a giant, anthropomorphized, robotic weapon)—all of these popular products provide ongoing access to a fascistic iconography in an acceptable (because purportedly nonpolitical) format. As Lorentzen remarks in his discussion of the persistence of "Reich style" in the image industries, "popular culture appears to provide realms of transgression whereby fascist iconography can be hijacked and repackaged according to prevailing tastes and fashion with only an echo of their Nazi context remaining" (1995, 161). These icons invariably present the male figure alone, standing tall, and in an attitude of concentration, Virgilian self-sacrifice, and preparedness. I speak here only of the iconography of these characters, for it is not unusual for the contemporary action-adventure film to have a core narrative that runs counter to the simplicity of its iconology. For instance, the narrative of *Rambo: First Blood, Part Two* has an anarchistic theme as a subtext to its overt patriotism, with the hero turning his wrath like a latter-day Luddite against the machines of the State and engaging in a "vengeful hunt" for cowardly and deceitful government officials (Molesworth 1986, 110). This is also true of the original Paul Verhoeven *RoboCop* (1987) film, with its obvious parallels to Lang's *Metropolis* (1926) (Verhoeven 1987, 33), its critique of hypermasculine society in the protagonist's symbolic rebirth (Taussig 1990), and in its explicit critique of the modern corporation and the "technologization of the human subject" (Wilson 1994, 295). And yet, the narratives have little or no semic control over the static icons they generate, and these often go on to have a life in their own in their manifestations as spin-off products.

This aesthetic resurfaces in some unexpected places throughout the range of the spectacle. The world of rock and roll, for instance, seems at a great remove from all this. The hirsute and anorexic appearance of the standard male rock stars circa 1968–80 and their assumed attendant politics (of either communalism, somewhat left of Democratic Party pedestrian liberalism, or anarchism, or perhaps nihilism) seem to have no connection with the political constellations described above. Reynolds and Price, however, demonstrate that many aspects of rock culture have an affinity with a kind of fascist aesthetic. In bands like Devo and Kraftwerk (an American and a German band, respectively, both reaching their height of visibility in the early 1980s), there is at work, in terms of sound (repetitious and machinelike music and vocal styles) and image (close-cropped hair, uniform appearance, soldierly stiffness), an informing "proto-fascist imagination" that is "riddled with envy of the machine, its invulnerability and impenetrability" (Reynolds and Press 1995, 102). While it is clear that in both these cases the intent was ironic (Devo with its atomic-plant protective gear and ziggurat-shaped hats; Kraftwerk with its image of mannequins at the control of computerized music systems, as on the cover of its *Computer World* album [1981]), we must, as noted in the case of the action-adventure image, remind ourselves that the iconic power of the image can operate autonomously of ironic intent. For example, in the case of another early 1980s band, DAF (Deutsch-Amerikanische Freundschaft), ironic and parodic intentions in reference to Germany's fascist past seem evident enough, particularly in their "Tanz den Adolf Hitler," which inspired a dance based on a caricature of the Nazi salute. The song's lyrics invite a comparison between popular culture's "dance crazes" and fascism: bend your knees, wiggle your hips, "tanz den Mussolini / dre dich nach rechts und mach den Adolf Hitler" (dance the Mussolini / then turn to the right and do the Adolf Hitler) (qtd. in Seeßlen 1994, 175). Nevertheless, the band's Naziesque image (as in the picture of the leather-clad DAF on the front cover of their *Gold und Liebe* album) may contain an atavistic appeal that cuts deeper than the irony (and even those who are well-aware of the irony may feel the emotional stir of fascism's will-to-power and "collective energy"). Interestingly enough, in 1992, about a decade after "Tanz den Adolf Hitler," the German government felt compelled to ban songs by five neo-Nazi rock groups ("Germany Acts" 1992). When an icon is presented in an ironic context, does the audience give itself to the irony, or to what Barthes calls "the primary virtue of the spectacle" (Barthes [1957] 1977, 15)? The presence of such parodic and satiric practices in popular culture has at root a complexity identified by Lorentzen:

> [O]n the one hand there is an element of genuine concern on the part of those writers who detect the insidious presence of fascist imagery in contemporary culture; on their other hand, there is the simultaneous admission that this "evil aesthetic" has a powerful attraction that appears to be beyond rational analysis. (1995, 162)

Another aspect of rock culture that shares these affinities can be found in the "heavy metal" genre, whose very name, as Robert Walser notes in his detailed study of the genre, conjures the armored body ideal of the Theweleitian soldier-male (1993, 116). This genre finds its impetus in founding bands like Steppenwolf, Led Zeppelin, and Black Sabbath, and in latter-day metal bands like Megadeth, Armored Saint, and Metallica, which in sound and image, claim Reynolds and Press, embody fascist worship, as in Marinetti's 1909 prefascist "Futurist Manifesto" (Marinetti 1972, 41–42), of violence, speed, and metal. Heavy metal is furthermore characterized by an imagistic fusion of technology (guitars, amplification, feedback, and distortion) and, as Walser notes, the archaic (as in the ample use of arbitrary umlauts in band names—Mötley Crüe, Queensrÿche—and the quotation of baroque melody in the music of Deep Purple and Yngwie Malmsteen), which brings us back to popular modernity as an antimodernist modernism. This pseudomedievalism is evident in other contemporary entertainments as well, particularly those inspired by what has been termed the neopagan movement: the structure of cyberspace's Multi-User Domains (or Dungeons) is derived from the "Dungeons and Dragons" game craze from the mid-1970s, which itself is a fusion of war-gaming with miniature figures and the medieval fantasies of J. R. R. Tolkien's novels, *Lord of the Rings* (1954) and *The Hobbit* (1937), which inspired a campus cult in the later 1960s (Holmes 1981, 12–13; cf. H. Carpenter 1977, 266). Furthermore, the heavy metal concert (and the heavy metal music videos that typically use concert footage) draws on the psychodynamics of the party rally discussed earlier; Walser, using terms similar to Bataille's in "The Psychological Structure of Fascism" notes the "collectivity and participation" (1993, 114) and a kind of national identity through the social clubs and dress styles (an antiestablishment uniform) of the "headbangers," as the fans call themselves. As far as gender is concerned: until the metal form became more amenable to a female audience (in the 1980s with bands like Van Halen and Bon Jovi), heavy metal was, and to a large extent still is, an imagistic venue for adolescent male identification; as Walser remarks, in heavy metal videos "women are presented as essentially mysterious and dangerous; they harm simply by being, for their attractiveness threatens to disrupt both male self-control and the collective strength of male bonding" (118). A similar aesthetic is operative in the related "punk" lineage that begins with Iggy Pop, is resurrected with the Sex Pistols, and is continued in a late 1990s "Gothic" performer like Marilyn Manson. While Iggy Pop's anarchistic image came with a palimpsest of militarism (particularly in a song like "Search and Destroy," and his pro-Reagan stance in the early 1980s), the Sex Pistols, more clearly identified with the anarchist left, intersect with all this through their Theweleitian dread of the female "abject" principle in both song (1977's "Bodies") and more palpably, in bass player Sid Vicious's horrific murder of his lover, Nancy Spungen (Reynolds and Press 1995, 85–95).

The image practice that is, or historically has been, controlled by men, consumed (watched) by men, and concerned primarily with women is that of pornography. Fascist literature, horror literature and film, and the aural/image texts of heavy metal music and video (some of which Tipper Gore referred to as "porn rock" [Walser 1993, 138]) constitute a parallel to pornography in that they are sites for the icono-graphic working out of our concerns about Self and Other, connec-tion and boundary. It is also evident in all of these masculinist image-discourses that the feminine principle is largely identified with the principle of disorder, often manifesting itself in terms of fecundity and fluidity.

Pornography has a long and complex history; its roots are found in the voyeuristic impulses in Ovid, the Song of Solomon, the *Kama Sutra*, and the graphic traditions of India and of the Far East. It was a component of religion and cultural life as part of an overall mythology of love and Eros. The erotic texts and images of popular modernity, however, are a different thing altogether; the word "pornography" itself did not make its first appearance in *Webster's* until 1864—a "trea-tise . . . on the subject of prostitutes." Definitions that we would more clearly identify with pornography, e.g., obscenity and the instilling of sexual excitation, did not make their way into standard dictionaries until much later (Hoff 1989, 39). The connection between pornogra-phy and prostitution (from the Greek, *pornos* + *graphi* = writing about whores) demonstrates its relationship to commerce, to what is regarded as morally reprehensible, and compels eroticism's "eventual dissociation from our larger cultural mythos" (Michelson 1971, 19). Pornography as an underground, subcultural image-discourse of the twentieth cen-tury has more immediate roots in the French Enlightenment—paral-leling the French Revolution, but more particularly, the writings of de Sade—and in the thriving pornographic industry of the Victorian period, as expressed graphically and in pornographic novels such as *My Secret Life* (1890) and *A Man with a Maid*. The origins of modern pornography in de Sade, who championed transgression in all its forms, and in Victorian England, the first society to fully experience industrialism and its social demands, indicate pornography's relationship to social control: pornography, that is, is much more than images of people engaged in sex acts. While modern pornography may be stripped of the mythological and religious significance associated with ancient erotica, the form and sequence of the sexual acts of modern visual pornography nevertheless engage a ritual that is simultaneously an actual act ("real sex") and a enactment of a culturally specific myth of social transgression—what Michelson calls a "mythos of animality" that contends that normal social deportment can be thrown over at any moment by powerful sexual urges, as in the standard porn narrative (Michelson 1971, 26). (This, incidentally, is paralleled by the "wolfman" genre in horror literature and film.) Further, as we think back to the beginning of this chapter and its sketch of the development of ocularcentrism and its relation to scientific and technological worldviews, the French Enlightenment and British Victorian roots of pornography become demystified, for while pornography is generally considered a peripheral and abnormal cultural discourse, we see that on a covert level sexual scopophilia reveals its kinship to scientific ocularcentrism.

The intersection between technology and the pornographic impulse occurred within the first decade of commercial photography—pornographic photographs were particularly popular among soldiers during the American Civil War. Similarly, pornographic motion pictures appeared soon after the technological process had been developed; and soon after the introduction of the VCR, videotape became the primary medium of pornography and the advantages of "audio visually simulated masturbation" in the privacy of one's home brought to a close the era of porn cinema (Marc 1984, 181). This process would be repeated with the introduction of interactive computer software—"virtual porn" or "interactive erotica"—about as soon as it was technologically feasible (cf. L. Williams 1999, 306–11). These are examples of what Tierney calls "the erotic technological impulse," which is rooted in what would appear to be the male love of technological gadgetry and pornography's essential nature; i.e., it is a substitute for some other "real" thing, and thus it has an ongoing recourse to technology in an effort to approximate the real (1994, 18).

By the 1920s pornography was abetted by the Fordist shift toward a producer *and* consumer society, which meant that "the emphasis in American life was shifting towards consumption, gratification, and pleasure" (D'Emilio and Freedman 1988, 172). Not coincidentally, Fordism itself was enabled by another visualist enterprise, motion pictures and the Hollywood system, which promoted the "triumph of an emancipatory consumer culture over the highly restrictive Victorian culture that dominated American life at the turn of the century" (S. J. Ross 1998, 176; cf. May 1980). The Fordist shift signals the onset of a consistent urging toward gratification and consumption. The sensualism that such a move entails has resulted in what Haug has called "a general sexualization of the human condition" and the growing tendency for consumer goods to "assume a sexual form to some extent" (1986, 55).

In addition to technological and cultural explanations, pornography—and modern pornography in particular—also requires a psychoanalytic explanation, particularly in terms of self-other relations and the ego boundary: in his essay on pornography and heterosexual male visual experience, Timothy Beneke uses Nancy Chodorow's theory of gender identification to propound a theory of heterosexual pornography that is more cognizant of male psychosexuality than either the notion of pornography as ersatz intercourse or the position of anti-porn feminists, most notably Andrea Dworkin, who argue that pornography in both its blatant and subtle forms serves to construct and support patriarchy, gynophobia, and the whole range of violent acts (from verbal abuse to domestic battery, rape, and murder) against women. Beneke points to the "loosening of one's identity boundaries" that is an inescapable element of sexual love, and he goes on to point out that this element is particularly threatening to the male because of how male identity is established. Men must make the passage from a primary mother-identification, established in birth and infancy through the mother-child dyad, to a father-identification; as the father is a more distant figure, the identification process is generally fraught with uncertainty and the interference of "grandiose cultural stereotypes of masculinity" (Beneke 1990, 179), which are themselves a manifestation of this tenuous process. As Beneke notes,

> Having sex with a woman threatens men's masculinity by unconsciously signaling a desire to regress and experience a certain infantile safety and union that was felt in the arms of the mother. Hence,

men defend themselves against this desire by relating to sex as an
occasion for domination. Pornography can be seen as a kind of training in the emotional work of achieving arousal and gratification
with minimal identification with women. (179)

It is thus that pornography, Beneke continues, replaces the more threatening (in terms of identification fears) tactile experience of sexuality
with a visual experience which provides a sensation of control and ego
autonomy; pornography is thus an iconographic system that, like the
fascist monument and the horror image, addresses the male anxiety
regarding self-boundary—the fear of being "absorbed." "Looking,"
says Beneke, "distances men from women; the stereotyped images of
porn further distance and defamiliarize women from men. Women
become the 'other'" (182).

In post-1970s hard-core heterosexual pornography, the relationship between soldier-male and dreaded female Other is exemplified
by that peculiar iconographic center of hard-core pornography, the
cum shot, in which the male performer withdraws his penis from the
female performer in order to ejaculate on her face or some other part
of her body. As a writer for *Penthouse* puts it, this image is the "focal
moment in porno . . . the equivalent of landing the killer punch,
breaking the finishing tape, or kissing the bride" (Haden-Guest 1997,
136). The significance of this metaphoric concatenation of destructive
violence, athletic victory, and sexual affection should not be lost on
us. While the cum shot can occasionally be found in pre-1970s pornography—an example from 1934 in the Kinsey collection (see the
bibliography) has all the markings of the cum shot icon, and, as Legman observes, there was a display of semen in 1960s pornography, in
Denmark in particular (1975, 133)—it was not a dominant image.
The pervasiveness of this image is suggested by the plethora of terms
that have slipped into contemporary vulgar slang: "facials," or "pearl
necklace," which is also the title of a 1981 song by rock band ZZ Top
(as the liner notes to ZZ Top's Greatest Hits [1992] tell us, the band
has its own ideas about "appropriate accessorizing. It's the band's in-
advertent contribution to relationship counselling"). These terms are
used repeatedly in the titles of pornographic magazines, videos, and
promotional websites, e.g., "Amazing Facials," "Nothin' But Facials,"
"Gangbang Cumshots," and, to give credence to the pornography-
horror connection, "Monster Facials"). Indeed, the image is so perva-

sive that it has spawned a counterimage: the "creampie" or "internal cum shot," in which semen is photographed as it drips from a vagina (Cream Pie 1999).

Linda Williams's analysis serves well to connect the concerns of both empirical and ocularcentric ideologies in her discussion of pornography's quest for visual "evidence." As pornography became more technologically sophisticated and more exposed to market competition, the simple evidence of penetration provided by the "meat shot" of the earlier stag film was not sufficient. As Williams explains, the cum shot is driven by a visual will-to-knowledge that asks viewers to believe that the performers "shift from a form of tactile to a form of visual pleasure at the crucial moment of the male's orgasm" (1999, 101). Hansen, Needham, and Nichols, in an essay comparing the discourse of pornography to the scientific discourse of ethnography, arrive at similar conclusions, noting that the cum shot is an "ironic documentation that requires violation of the 'real' sexuality it represents," offering "measurable proof of the inward and subjective state of ecstasy that would have been attained had the need of the observer not received priority" (1989, 74). In a recent development, this desire to provide the "evidence" of male sexual pleasure has given way to "natural magic" and illusory spectatorship as many pornographic videos, and particularly the cover photographs that entice one to buy, feature partially concealed artificial yet very realistic phalluses that shoot decidedly unrealistic quantities of artificial semen when squeezed, and so the titillation of seeing "real" sex and "real" semen has in this particular variation given over to illusionism.

The formula of the cum shot cinemagraphic sequence generally demands that after the moment of ejaculation the woman gazes directly into the camera; this brings to mind Berger's observation that the female subject in the Western tradition of nude painting (he uses Ingres as an example) looks out of the frame at her owner, at the male viewer. Additionally, during this sequence the female performer will often directly address the viewer ("I want you to cum in my mouth"), thus completing the identification fantasy that allows the viewer to fictively replace the male performer. However, this is not merely a carnal fantasy; it is also an emotional one—a fantasy of "unconditional acceptance" in which the female seems to say "I exist wholly for you. I will never reject you. You *cannot* disappoint me" (Bordo 1993, 707). With the camera centered on the female performer, she will in many

videos "dirty talk" to either the male performer or, in a psychologically knowing variation, to the camera in direct address to the viewer.
For the masturbating male viewer, the camera occupies the same imaginative space as his penis, and it is worth noting the semantic connections between the entities herein involved: the "cum shot" refers to
both a camera "shot" and the actual "shooting" of the ejaculate. We
speak of "loading a camera," "shooting a camera," and "shooting a
load." There are also examples of the opening of the penis as an imaginative eye, as in the vulgarism, "one-eyed trouser snake," or in Molly
Bloom's soliloquy in James Joyce's *Ulysses* (whose publication in the
United States was delayed until 1934 following a 1933 Supreme Court
ruling that lifted the ban on this "obscene" novel), in which she (or
rather, the male author trying to imagine a female perspective) speaks
of her first sexual encounter in which she "unbuttoned him and took
him out and drew back the skin it had a kind of eye in it" (Joyce
[1922] 1961, 760–61), or in the concatenation of penis-camera-weapon
in the groundbreaking horror film, Michael Powell's *Peeping Tom* (1960)
(cf. Clover 1991, 168–80).

   In what is generally the last scene of these rule-governed performances, the female performer displays as conspicuously as possible the
semen that has been deposited on her face, sometimes rubbing it on
her lips, rolling it with her tongue, letting it drip from her chin. This
is a spectacle of horrific femininity for the male imagination—attrac-
tive and repulsive at the same time. It is the horrific spectacle of
Lovecraft's "Lurking Fear": "Seething, stewing, surging, bubbling like
serpents' slime it rolled up and out of that yawning hole, spreading
like a septic contagion" ([1925] 1971b, 20). In another Lovecraft tale,
"From Beyond," a mad scientist creates a machine that allows the
spectator to see a hidden world. As the narrator explains, the impression created by the machine was "familiar, for the unusual part was super-
imposed upon the usual terrestrial scene much as a cinema view may
be thrown upon the painted curtain of a theater." In this setting, as in
a darkened theater, the narrator sees a horrible biological otherworld:

> [F]oremost among the living objects were inky, jellyfish monstrosi-
> ties which flabbily quivered in harmony with the vibrations from
> the machine. They were present in loathsome profusion, and I saw
> to my horror that they *overlapped*, that they were semi-fluid. . . .
> Sometimes they appeared to devour one another. ([1920] 1971a,
> 64–65)

This cinematic convention of the cum shot, then, is informed by both an ocularcentric ideology and the Chodorowian notion of the male's fear of identifying with the female, for at the crucial moment of biological merger the male performer backs away and breaks physical contact as his fluid flies onto the female performer's face and lips. In terms of masculinist iconography, the feminine principle of fluidity separates from the male principle of solidity ("hardness"—which has masculinist moral and military associations, as its use in the Jünger passage cited earlier) and goes to "where it belongs": with the woman. Thus, the entire framing of the shot has a binary arrangement in which one side of the frame contains all that is, according to a sexist ideology, feminine (woman, kneeling before the master in worship of the fetish object, neck bent [a general sign of submission in human and animal behavior], fluidity, the open mouth), while the opposing side of the frame consists of idealized masculinity (standing, but with the head often cut from the frame; his erect penis has replaced his face just as the woman's mouth has replaced her vagina, and indeed, in pornography, man *is* both penis and phallus, for here the symbolic and the real merge [cf. L. Williams 1989, 209]). Thus the scene enacts a symbolic separation of idealized masculinity from what Kristeva calls the "abjection" of women and what Theweleit identifies as the male fear of the female/fluid principle—the viscous fluidity of breast milk, semen, vaginal secretions, saliva, pus, urine, all of which "blur the border between me and not-me, inside and out. From the soldier-male perspective, these things are neither liquid nor solid but *sullied*; slime that must be expelled" (Reynolds and Press 1995, 85).

Finally, the sexism implied here is evident throughout the icon, through the implied domination of the standing male/kneeling woman stance that is one of the most popular forms of this image, but also in the implied degradation of the phallus "spitting" on the woman's face—that part of the body which is most closely associated with one's individual dignity and personality—thus turning the object of desire into an object of scorn and providing "visual proof of her objectification and humiliation" (L. Williams 1999, 112). (According to the protofascist Marinetti, women are a kind of "poison" and are to be regarded as objects of "scorn" [{1909} 1972, 42, 72].) It is thus an act that combines sexual cathexis of man for woman, the pornotopic fantasy of the totally yielding and accepting female, with, strangely, aggression toward women for failing to yield universally to male desire and likewise punishing the object for its power over the passions of the dominator. As Griffin perceptively notes, the failure to acknowledge the power of nature and eros can be "expressed as a fear and hatred of these forces. Rather than acknowledge eros as a part of its own being, this cast of mind divide[s] itself from nature" and then expresses a "fear of the power of nature as nature returned to its consciousness in the body of women, or 'the Jew,' or 'darkness'" (1981, 11). An aspect of this fear is found in Sartre, who, in *Being and Nothingness* (1943), described a woman's vagina as a "voracious mouth which devours the penis" in an act not unlike castration (qtd. in Collins and Pierce 1976, 118). The cum shot, then, compounds the "safe distancing" of the woman that is general to pornographic representations with the added "safety" of freeing the penis from the potentially castrating female orifices.

Pornography has generally been regarded as a marginal image-discourse—too marginal to play a central role in the image regime of popular modernity. However, because of the inherent sensualism of commodity exchange (Haug 1986, 17), and more specifically because of the proliferation of image-technologies—from the daguerreotype to the VCR and the Internet—what had been the "secret museum" of pornography is now available to virtually everyone. In 1996 the porno industry was estimated to generate an income of roughly seven billion dollars—hardly a marginal practice at the cultural periphery (Slade 1997, 1–12). Indeed, the example of modern pornography gives us ample reason to rethink the concept of center and periphery in culture. Is the periphery the wellspring of the center—the roots that nourish the center? Perhaps more significant than de facto pornography is its assimilated presence in the broader commercial culture; as Michelson argues, much of the pornographic genre has become part of "the mainstream of cultural life" (1971, 19) in what we might call the pornographization of popular modernity, a development rooted in the Fordist shift toward self-gratification and mapped by a lineage of entertainments and entertainers who have blurred the line between the mainstream and the marginal: Mae West, Hugh Heffner and his "Playboy Philosophy," Jayne Mansfield, the James Bond films of the 1960s, and, as we'll discuss shortly, music videos in the 1980s. We can also note the strong interplay between trends in mainstream entertainment and pornography as pornography has broken out of its "billion-dollar ghetto" (Kroll 1999, 70); for example, just as nostalgia, as we noted previously, has become a dominant element in the mainstream entertainment industry, so too has it become a part of pornography: in the 1990s, pornographers began to market nostalgia porn in the form of re-released "classics" (with performers like John Holmes, Seka, and Vanessa del Rio) from the early 1970s, as websites like "Retrosex.Com" (1999) exploited a taste for black-and-white "stag" stills from the 1950s.

One salient example of the increasing eroticism of public imagery can be found in the veiled pornography employed in the marketing of nonessential products that already have a quasi-sexual appeal through their primary oralism, such as beverages and cigarettes. Billboard and magazine advertisements for Newport cigarettes (Lorillard, Inc.) have been particularly noteworthy in this regard; Moog observes that Newport roots its advertising success in their ability to associate cigarettes (drawing on the same mechanism described by R. Williams as a "magic system" [1993, 320] "with themes of sexual dominance and submission"; they use the "Alive with Pleasure!" ads to suggest that "its smokers will become *sexually* alive with pleasure." One ad that Moog analyzes shows two men carrying a woman on a long pole in "deer bounty fashion" while their spatial positions strongly suggest a ménage à trois (Moog 1990, 148). Another ad from about 1975 featured a grinning male with one arm around a female while with his other hand he squirts her in the face with a garden hose. It is interesting to note that this particular image was formulated just after mainstreamed hard-core pornographic films (e.g., *Deep Throat* [1972]; *The Devil in Miss Jones* [1973]) and increasing access to hard-core had deposited the informing iconographic structure into general public culture.

Another good example of this iconography is found in ads for soft drinks, and as the evolution of this form of advertising, and indeed, the product itself, reveals much about both American popular iconography and popular modernity, a brief overview is in order. In the first chapter, I argued that the dissemination of telephony marks "the moment" of popular modernity because it brought technology into the domestic sphere in an unprecedented way, reformed social relations, and ultimately, led to a redefinition of consumer-producer relations. But we should also note that these developments were prefigured in the change in Western nutritional habits during the emergence of modernity in the broadest historical sense. As Stanley Mintz (1985) points out, few Europeans knew of the existence of sugarcane and its processed extract (table sugar, sucrose) before the year 1000 A.D.; it became a staple among the English nobility by 1650; by 1800 it was a regular, though costly, element of the common English diet; and by 1900 it supplied one-fifth the calories in the average Englishman's diet. By this we see that modernity colonized the physical body of its subjects long before the technological colonization discussed in various ways in this book. In the late seventeenth and eighteenth centuries, sugar developed its long-standing relationships with other colonial stimulants (chocolate, tea, coffee), and the relationship between industrial labor and the consumption of stimulants, as in the British working class's long-standing demand for tea and sugar. Indeed, the caffeine-sucrose combination, virtually unknown before the mid-seventeenth century, became, during the nineteenth century, commonplace. One might call it the beverage of popular modernity, and one might speculate on the way in which this stimulant combination (we could add tobacco to the list) creates the biological conditions necessary for modernity. Likewise, Pinkus notes the relationship between sugar, modernity, and fascism with specific reference to the Italian futurists "banishment" of pasta from the diet in favor of sugar and stimulants, all for the purpose of eliminating Italian backwardness and sluggishness (associated with pasta) in favor of apparently "modern" nervous energy (1995, 88–89).

We know that sugar (refined sucrose from the sugarcane plant) was introduced to Europeans by the Arabs. Drawing to some extent on classical sources, Arab pharmacology incorporated sucrose as a medicinal substance, and these ideas came to Europe by way of Spain as early as the ninth century. One of its pharmacological uses was that of making "elixirs," often for the purpose of aiding digestion (Mintz 1985, 96–100). We should not be surprised, then, that the soft drinks that are a fixed lifestyle feature of popular modernity were invariably created by pharmacists and were alleged to have therapeutic value. Root beer, a fermented, low-alcohol content sarsaparilla and sassafras beverage developed by Philadelphia pharmacist Charles Hires in the mid-nineteenth century, was superseded by the end of the century by beverages that used the extract of kola nuts (grown in Africa and South America, and long used in these places as a mild stimulant, and in some cases, an aphrodisiac) in combination with sucrose and carbonated water. This is, of course, the basic formula for both Coca-Cola (which originally employed cocaine as well) and Pepsi-Cola, and all other cola-based soft drinks.

Curiously, the establishment of soft drinks in the popular-modern lifestyle is implicated in both the older medicinal tradition of sucrose and in the distinctly modern phenomenon that Lears calls the "therapeutic ethos," an ideological structure that was a result of the disruptions of urbanization, industrialization, and mass consumption at the end of the nineteenth century, disruptions that resulted in feelings of "unreality" and anomie among the urban bourgeoisie. The therapeutic ethos, fueled by these changes and by a concurrent nostalgia "for the vigorous health allegedly enjoyed by farmers, children, and others 'close to nature'" (1983, 11), had many cultural and commercial manifestations. Culturally, there were numerous works that called for self-renewal. Vitalist manifestos to youth, vitality, and enthusiasm appeared in the writings of psychologist and educationalist G. Stanley Hall (whose critique of geography education we noted earlier); and commercially, these perceptions and ideologies were harnessed by the new, psychologically sophisticated advertising of the 1920s (inspired by Edward Bernays, who we'll return to soon), which consistently aroused consumer demand by "associating products with imaginary states of well-being" (Lears 1983, 19).

And there were also products, including soft drinks, which are in fact a distant echo of the mythic reinvigorating elixirs of the ancient physician, the Arab pharmacologist, and the medieval physician. Pepsi was invented by Caleb Bradham in 1898 in a Bern, South Carolina drugstore operation, and it was originally promoted, true to its heritage, as both a stimulant and a medicine; as the 1903 promotional rhetoric put it, "exhilarating, invigorating, aids digestion." By the 1920s, these drugstore operations evolved into national business concerns, in much the same way that tobacco moved toward national brand markets. During the Depression, Pepsi ran a very poor second to Coca-Cola; when Walter Mack took the company over in 1933, he soon turned to a relatively new medium—radio—and developed what he claims to be the first radio "jingle." Using a traditional folk melody (the British ballad, "D'ye Ken John Peel"), Mack's jingle appealed primarily to the consumer's budget during these lean years (Pepsi Generation Collection 1984 [Mack]). The product was marketed in twelve-ounce bottles (as opposed to Coca-Cola's six) at the same price of five cents: "twice as much for a nickel," as the jingle tells us. Pepsi ads from the forties tended to stress the war effort: as one ad put it, "You're pouring a great drink there, mister. Great in quick food energy that's vital to you and America." It was during this time that the red, white, and blue colors of the Pepsi icon were introduced as a sign of corporate patriotism, thus continuing a practice from at least the first decade of the century, that of using nationalist imagery, as discussed earlier in reference to the new national brand tobacco markets (fig. 1). After the war, the focus returned to Mack's price-competition emphasis.

In the 1950s the company came under the leadership of Albert Steele and the influence of the Kenyon and Eckhart advertising agency, and not insignificantly, a tie-in with Hollywood through Steele's wife, Joan Crawford. Pepsi, largely excluded from the fountain market because of Coca-Cola's established dominance there, decided to pursue the home market. In the economic climate of the 1950s, Mack's price-competition formula had become irrelevant, and as postwar American suburbanites wanted an identity that was in keeping with the new prosperity, Pepsi's reputation as a cheap alternative became a handicap. There was even a racial element involved: Pepsi's lower price gave it dominance in black retail markets, in which consumers had far fewer dollars for luxury spending; this led Pepsi to be known, especially in the South, as "the nigger Coke."

The Pepsi campaigns under Kenyon and Eckhart made the decisive leap from price competition to image competition. Knowing that women were the primary purchasers in the grocery outlet market, Pepsi lowered the sugar content and marketed itself with reference to female body image: the John Peel folk song, which had become a priced-based advertisement, now was reformed into an appeal to body image and modernity. As Polly Bergen crooned in Pepsi television commercials in the 1950s, "Pepsi-Cola tastes just great / For modern folks who watch their weight." At about the same time, a television and magazine campaign entitled "The Sociables" was launched: slogans like "the light refreshment" and "be sociable, have a Pepsi" were combined with decidedly suburban, modern, middle-class and even haute bourgeois imagery ("they entertain the modern way"). Subtle sexual imagery was also employed, with men casting admiring glances at young women with slim waists, and with the newly redesigned bottle (with curves and a twirl design known in the bottling trade as a "skirt") playing a prominent role.

But the Kenyon and Eckhart/Joan Crawford phase was an intermediary one. While the appeal of a lower sugar content than Coca-Cola (previously, Pepsi's sugar content had been actually higher than

Coke's) was effective, the appeal to class snobbery (some ads even featured elegant evening clothes) was judged a failure; in its attempt to dispel its working-class identity, Pepsi had gone too far. In the early 1960s, Pepsi moved its account to the advertising giant BBD&O (Batton, Barton, Durstine, and Osburn).

As revealed in the many interviews with BBD&O and Pepsi employees and executives who were involved in the advertising campaigns of the 1960s ("come alive"; "think young"; "Pepsi Generation"), both the price competition strategy and the snob appeal ads were decisively rejected (Pepsi Generation 1984); this brings us once again to our consumerism/fascism comparison, for as Pinkus notes, advertising during Italy's fascist period sought to create a "classless consumer who would inevitably identify his or her needs with a national economy rather than with a particular class-based commodity culture" (1995, 82–83). The election of America's youngest president, John F. Kennedy, in 1960 became the informing event for the new television ads during that decade, which consistently stressed youth, collective energy (athletic activity on California beaches became a major setting, and the cinemagraphic style was decidedly kinetic), and leisure without specific reference to social class. In short, the informing elements of the campaign revived elements of the fascist aesthetic. This is perhaps best seen in the television commercial that defined the new Pepsi advertising style. The commercial, entitled "Rope Swing" (1968), featured athletic young bodies diving into a creek using the backlighting method for dramatic effect—one of Riefenstahl's signature techniques that had been largely forbidden by the neutral aesthetic of the American televisual style. Ed Vorkapich, who directed the commercial, took his inspiration from the famous "divers" scene from the film that largely defined the fascist aesthetic in all its facets, Riefenstahl's *Triumph of the Will* (1935), a film he came to admire greatly through his formal study of cinemagraphic history (Pepsi Generation Collection 1984 [Lipsitz]). We can view this as an insignificant coincidence, a mere borrowing of dramatic pictorial techniques—but I would contend that such aesthetic borrowings cannot be made without carrying traces of their original ideological content. Further, as noted earlier, both fascism and consumerism are strategies for managing mass populations in the context of popular modernity, and given this, it should not surprise us that they generate congruent (rather than discrete) image-discourses. One of the more interesting connections in this regard is found in an aspect of the influence of Edward Bernays, the nephew of Sigmund Freud who played a major role in the formation of tobacco advertising and virtually all forms of advertising and public relations since the 1920s. His magnum opus, *Crystallizing Public Opinion* (1923), became a major source for the Nazi propaganda chief Joseph Goebbels in his engineering of a public opinion campaign aimed against German Jews. (Tye 1998, 111).

In tracing these other strands that emanate from the skein of popular-modern iconography, we've temporarily lost sight of our tracing of the pornographic. While the Pepsi ads from the 1950s used sexual imagery, and the BBD&O campaigns continued this usage (young attractive women; a new theme song, "For Those Who Think Young," that promises to defeat time, reminding us of the role of chronophobia in this cultural nexus, and simultaneously has a sexual tinge, as the song was sung by a winsome female voice, that of Joanie Summers, to the tune of "Makin' Whoopee"), in the post-1970s period there was an intensification of this approach. One Pepsi television commercial from the 1980s illustrates well the tight relationship between the various image strands of masculinist popular modernity. The scene is that of a rock band rehearsal; a ice-cold case of Pepsi is brought to the male band members and their female entourage; unfortunately, they don't have a bottle opener. The lead guitarist (and we'll turn to the sexual iconography of the electric guitar later) then performs a frenzied solo, his gaze focused on the Pepsi bottles. His frenzied performance is cross-cut with the image of excited female faces; soon the bottles are foaming and about to blow off their caps, and finally they do as he brings the solo to a piercing crescendo (Pepsi-Cola Television Commercials 1985). In another Pepsi commercial from the same period, model Cindy Crawford pulls her cherry-red sports car into a rural gas station. Two young boys spy on her from a concealed wooded area as Crawford gets out of the car and makes her way to the Pepsi machine. The cross-cutting here is between the two young voyeurs (the nostalgic background music is provided by a popular "classic" from the 1950s— "Just One Look") and Crawford's face and lips as she brings the bottle to her lips in a passionate gesture (the framing of her upper torso makes her white T-shirt look like an undergarment, and the sign in the gas station window behind her reads "Open"). The punch line, as it were, of the ad is that the boys, in true adolescent self-absorption, are admiring the bottle, not Cindy (Pepsi-Cola Television Commercials 1988). The constant spilling of fluids in contemporary pornography and advertisements is paralleled by the increased use of artificial blood in horror films. In the late 1950s, Hammer Studios (part of the British initiative to capture a larger share of the North American entertainment market, as mentioned previously) took the lead in the horror film industry by featuring much more blood, especially in *Dracula* (1958), starring Christopher Lee, than in any previous horror movies. Following the trend established by Hammer, and more directly, the work of Hershel Gordon Lewis *(Blood Feast* [1963]), directors in the early 1970s such as Tobe Hooper *(Texas Chain Saw Massacre* [1974]), George Romero *(Night of the Living Dead* [1968]), and Wes Craven *(Last House on the Left* [1972]) brought more gory cinematic practices into the mainstream in genres that came to be known as "slasher" and "splatter."

Pornography is to spectacular sexuality what slasher is to spectacular violence. As Carol Clover has remarked, the slasher genre is "drenched in taboo and encroaching vigorously on the pornographic . . . beyond the purview of the respectable (middle-aged, middle-class) audience" (1991, 21). Just as pornography must show the liquid "evidence" of sexual pleasure; the slasher films like *The Texas Chain Saw Massacre* insist on lavish displays of blood, often smeared and splattered on the victim's body and face. This concatenation of semen and blood was early evident in the novel *Dracula* in the scene we already referenced, in which one of the characters remarks, "[N]o man knows, till he experiences it, what it is to feel his own life-blood drawn away into the veins of the woman he loves" (Stoker 1978, 137). Moving beyond the veiled imagery of this Victorian novel, the murders of the slasher film, as Clover observes, are "blatantly phallic" (1991, 42), the weapons of choice generally being chain saws, power drills, and the like. The perpetrators are markedly scopophilic, as is the very cinematic construction of the narratives themselves, which employ what Clover calls the "eye of horror." The "crudity and compulsive repetitiveness" (22) of slasher also links it to pornography. (A film that recognizes this connection most explicitly is *The Howling* [1981], which begins with a man turning into a werewolf as he watches a sadistic porno film in a viewing booth in a sleazy Los Angeles adult bookstore.) Another obvious connection between slasher and pornography is that the object of these cinematic assaults is the image of woman; arguably, this is related to the rise of feminism, the new female sexual freedom afforded by improved birth control technology, male unemployment resulting from de-industrialization, and the movement of women into new sectors of the workforce. Indeed, as Clover lucidly argues, inasmuch as slasher presents women as objects to be stalked and destroyed, the genre also presents women as being capable of defending themselves and repelling the male attacker, as witness Jamie Lee Curtis's definitive role in *Halloween* (1978).

Finally, like the pornography it parallels, slasher in the 1970s and early eighties was generally regarded as being at the margins of mass culture in spite of the fact that the financial evidence indicated otherwise: *Halloween,* for instance, produced for the mere sum of $320,000, had grossed a total of $75 million after six years (Clover 1991, 23). Another similarity is that both hard-core pornography and slasher have been considerably mainstreamed, moving from, in the case of pornography, old theaters in urban tenderloin districts to suburban video outlets, and in the case of slasher, from run-down drive-ins to state-of-the-art multiplex cinemas. More recent developments, most notably Clive Barker's *Hellraiser* films (a series begun in 1987), are characterized by a sadomasochistic sexuality: the creatures of hell often have a phallic appearance and are usually coated with plenty of slime, while the symbol of evil in the series, the devilish "Pinhead," rules over a Sadian heterotopia of pain and pleasure.

The male image has been transmuted in the general pornographization of popular modernity's image nexus, starting first with another marginalized image-discourse—bodybuilding. The rise in the popularity of the bodybuilder image is but another manifestation of the inevitable law of the ocularcentric society: just as pornography's development was driven by a will-to-see that eventually culminated (climaxed?) in the cum shot, in which, as Linda Williams stated, the reality of orgasmic pleasure is reduced to pure surface and visuality, the bodybuilder turns the ideal of the beautiful male body (a concept, as Mosse argues in *The Image of Man*, rooted in the classicism of Winkelmann and reaching its apogee in the aesthetics of fascism) into a frenzy of physiological revelation. The bodybuilder shaves his body, which has been starved of even the thinnest layer of subcutaneous fat. By the

1980s, this imperative had been taken to even more dangerous levels as competitors took to the technique of injecting subcutaneous diuretics to make the skin so translucent that, as Sam Fussell eloquently puts it, one can see "raw tissue and striated muscle swimming in a bowl of veins beneath" (1993, 583). The physical strain of this competitive sport promotes high blood-pressure, organ damage, and a generally shorter life span, as was brought to public attention in the 1996 death of the champion bodybuilder Andreas Munzer. The physical strain of bodybuilding stands as a metaphor for the spectacle society and its sacrifice of the internal for the external, for the level of sacrifice it involves suggests a kind of cultic civil religion. The ultimate semiotic impulse of bodybuilding reveals another aspect of its kinship with pornography, for the slickly oiled bodybuilder, as he stands on display with the appearance of almost painful hardness, is a living metonym for the erect penis. The bodybuilder, says Fussell, is, "quite literally, the cock of the walk" (580); he is an overdetermined image—he is both the spurting phallus/penis of pornography and the armor-clad soldier-male of fascist iconography, and his emergence from subculture to popular culture indicates the complicity of these two iconographic discourses.

A similar construction is found in one of rock's central and oft-repeated icons: that of the guitar player and the electric guitar itself. Leo Fender and Les Paul's experiments in the 1940s with electrically amplified, solid-body guitars remained a marginal feature in popular music until these innovations were commercially exploited in the early fifties and they became the defining element of rock and roll, displacing the brass of the big band era and the piano from the ragtime and New Orleans jazz traditions. The electric guitar allowed a new freedom of movement for the performer, beginning with the acrobatics of T-Bone Walker and Chuck Berry. Eventually the movement became part of a relatively stable stock of gestures, which in turn became part of popular iconography through images primarily taken from concert footage.

We have to question whether the guitar became the instrument of rock for its aural or for its iconographic properties, for the image of the rock star is almost unimaginable without the guitar, and, unmistakably, it is the male performer who is associated with this instrument. The guitar is also the focal point of identity for the male viewer, particularly for the adolescent male. The latter often identifies with the performer by playing "air guitar," a distinctly male response rarely taken up by female fans (Bayton 1997, 43). As Mavis Bayton points out, female electric guitar players are seldom featured or depicted in either the articles or the advertising of popular music magazines. There are a number of reasons for this. First of all, there is the masculinist-technology connection we've noted before; in this instance of that connection, the technological ambience of the electric guitar makes it as much a machine as a musical instrument (already here we see a male:female binary), and thus, as Bayton notes, it is an instrument that women are not likely to gravitate toward, given the ideological conditioning of patriarchy. The phallicism of the guitar also extends to its timbral quality: the sheer volume and attack of the instrument, says Bayton, "connotes phallic power" (43).

But most important, to return to the iconographic question, is the instrument's shape. The electric guitar (instrument:machine) is sometimes like a woman's torso (whose shape is caressed by the male performer), but more often, it is given a phallic (recall our Pepsi commercial) rendering. The image, and the identification of with the instrument through "air guitar" mimicry, is another instance of Ihde's notion of the embodiment relation, in which user and technology become, phenomenologically speaking, one. As Bayton remarks, echoing the Heideggerian-Ihdeian concept of the embodiment relation, "[T]he electric guitar, as situated within the masculinist discourse of rock, is virtually seen as an extension of the male body . . . with legs firmly planted akimbo, the guitarist is able to lean back in a parody of sexual ecstasy" (43). And like the phallus (as used by the rapist), the electric guitar can be, as Pete Townsend (guitarist for The Who) pointed out, "a weapon, above everything else" (Townsend 1995). An interesting image conflation in this regard is found in the slasher film, *Slumber Party Massacre II* (1987), in which the psycho-killer's weapon of choice is an electric guitar that has been outfitted with a power drill.

In the plethora of concert photographs of Jimi Hendrix, the musician who more than anyone else expanded this iconographic stock, we see clearly the entire polyvalent range of guitar symbolism: an advanced technological weapon (generating the war-soundscape of his Woodstock version of "The Star Spangled Banner"), a woman's vulva (as he "performs cunnilingus" on the strings while clutching the guitar's "hips"), or his phallus. (Prince—a.k.a. "The Artist Formerly Known as Prince"—later pushed the porno-music-phallus-technology-illusionism boundary by devising up with a guitar that shoots artificial ejaculate [Bayton 1997, 43].) And finally, Hendrix used his instrument to create an icon of quasi-religious sacrifice, as in a well-known image of him, captured in D. A. Pennebaker's documentary film of the 1967 Monterey International Pop Festival, in which the kneels before his instrument as it burns.

Another major figure in this iconography is Jimmy Page (Led Zeppelin), whose characteristic pose—shirt open and chest exposed, guitar held at crotch level, face bent forward and hidden by flowing locks of hair—became an icon-template for any number of his contemporaries (Peter Frampton, Freddie Mercury, Ted Nugent, Joe Perry of Arrowsmith) and imitators in the next generation of rock (Slash, of Guns 'N' Roses). The protofascist imagery of Page and Led Zeppelin has been noted: Reynolds and Press (1995) point to the Viking invader fantasy in their "Immigrant Song" (1970), and Davis, in *Hammer of the Gods*, points out that "Whole Lotta Love," with its Hendrixesque weapon noises, became a war anthem for American troops in Vietnam, who drove their tanks into battle with eight-track cartridge players blaring this sex and war anthem (1985, 102). This image nexus is, though, even more complex than the masculinist conflation of imagery can explain, for it combines fascistic masculinity (hard, machine-like—Marinetti would have approved) with feminized masculinity (Christlike—thin body, long hair, head hanging, expressions of pain and posture of sacrifice—Marinetti would *not* have approved). With the guitar—which is shaped like a woman's hips and sometimes suggests the idealized shape of the human heart—poised before the open shirt, the image is sometimes weirdly evocative of the Catholic icon of the Sacred Heart (cf. M. P. Carroll 1989; M. T. Carroll 1993). This indicates, perhaps, as Mosse suggests in the final chapter of *The Image of Man*, that the late twentieth century has seen a softening of the male ideal (much to the concern of Robert Bly, who, in *Iron John*, probes the problems of manhood and its deformation since the Industrial Revolution). It also suggests that all of these image discourses, strangely enough, border on religious imagery that harks back to the Greek Orthodox root meaning of "icon." If we think of the possible functions of popular iconography—stimulating consumption, abetting capital formation, pacification, pleasure, propagating a national mythos—we might also remember that, as Miles puts it, the primary task of any culture is a semiotic one: "to formulate and make available to its members effective symbols, whether verbal, visual, or a combination of the two, for comprehending and taking an attitude toward bodily experience: birth, growth, maturation, kinship, sex, life cycle, pain, death" (1985, 82). Religion as "a prominent aspect of culture" was until very recently, the primary institution for governing this process of semiosis (82). And as we've reminded ourselves of the standard use of the term "icon," we might also dwell a moment on the dual use of the term "mass," as in "mass culture" and the Catholic Sunday gathering.

Weaver has pointed to the connection between pornography and the religious imagination in the writings of John of the Cross, and more dramatically in the confessions of Teresa of Avila, whose confessions form a kind of spiritual pornography, or perhaps pornographic spirituality:

> The Lord wanted me while in this state to see sometimes the following vision: I saw close to me toward my left side an angel in bodily form . . . I saw in his hands a large golden dart and at the end of the iron tip there appeared to be a little fire. It seemed to me this angel plunged the dart several times into my heart and that it reached deep within me. When he drew it out, I thought he was carrying off with him the deepest part of me; and he left me all on fire with great love of God. The pain was so great that it made me moan, and the sweetness this greatest pain caused me was so superabundant that there is no desire capable of taking it away. (Qtd. in Weaver 1989, 77)

This erotic and transcendent moment (the moment of the "transverberation" of her heart by the Lord) was so powerful a stimulus to the Catholic imagination that in 1726 Pope Benedict XIII decreed a festival in its commemoration, and it became one of the most prominent Catholic icons owing to its baroque rendering in the famous Bernini statue. The face of Bernini's Teresa is virtually indistinguishable from images of women's faces in pornography, and both are similar to the faces of women about to be bitten by the vampire in any number of popular films. The connection between these images is theoretically supported by Ernst Jones's study of the roots of the vampire myth in wish-fulfillment sexual fantasies and in Rudolph Otto's of the numinous religious experience, which is akin to a kind of sublime fear (Otto 1928, 13; Ernst Jones 1931, 111; cf. N. Carroll 1990, 165, 170). These connections demonstrate that the culture of the baroque is truly the preindustrial precursor to contemporary popular culture (cf. Maraval 1986), a supposition made all the more credible by Pardun and McKee's study of religious symbolism in MTV music videos. These videos often have the effect of being a collage of images rather than a flowing narrative as in traditional Hollywood style; thus, they are not so much "motion pictures" as they are a series of relatively static images, making them the perfect locus for a contemporary iconography. In heavy metal videos from bands like Van Halen (with David Lee Roth), Kaplan observes an intense and adolescent identification with the phallus that goes beyond the Hollywood voyeuristic gaze through "zoom shots [of] male crotches and bare chests" and aggressive facial expressions (1987, 93, 102). But it is also true that in music videos featuring female stars (Salt 'N' Pepa, Madonna) the images often constitute a quasi-pornography. The combination of this sexual imagery with sacred symbology is found in videos featuring performers like Black Sabbath, Judas Priest, Ministry, The Church, and particularly Madonna. This transgressive impulse is so prevalent that Pardun and McKee's study demonstrates through quantification that religious images are more likely to occur in these videos in the presence of sexual images than without them (1995, 440, 445).

Transgression, Transcendence, and the Boundary

In the first section of this chapter, I maintained that popular modernity's image nexus is a site of subliminal struggle, for the negotiation of the boundaries between Self and Other. Fascist and masculinist images and icons are generally invested in maintaining that boundary and rejecting the heteronomous. To this end, the Nazis were the self-proclaimed "guardians of respectability," who banned all forms of erotic literature and pictures soon after coming to power, a move that gave them the overwhelming approval of the influential German Purity Leagues (Mosse 1996b, 175). Arguably, the marginalized military culture of the Freikorpsmen later became mainstreamed into the dominant culture of Nazism, whose bourgeois sentimentalism preserved images of order in every form, down to the most minuscule example of Nazi kitsch, while suppressing those of disorder, or rather, ghettoizing them in the activities of the death camps and the battle front.

But the desire for self-definition must be understood as being paired with its opposite: the desire to cross boundaries. The erotic charge many get from pornography probably has far more to do with the desire to transgress conventional social boundaries than with biological urges (cf. Bataille 1977, 29), an assertion that becomes all the more warranted when one considers some of the titles of contemporary porno novels: *Teacher's Naughty Lessons, Horny Foster Mother, Horny Cheerleaders*, to name just a few. These titles, not surprisingly, correspond to narratives of transgression—incest, pederasty, the abuse of social position, and so on. In the French tradition, unlike the American, pornographic stories and photos often draw on blasphemy, involving clerical characters (priests and nuns) or props of Roman Catholicism (e.g., holy water founts). The fact that such images loom large in French and not American pornography seems to demonstrate that eroticism's charge is rooted in transgression, which in turn is defined by a given culture's structure of authority, a structure that, in France, is strongly tied to Catholicism. Interesting in this regard are a set of pornographic photos produced in France circa 1915, which include a priest blessing a group of masturbating nuns and an altar boy sodomizing a priest (Kinsey Institute c. 1934).

The foundational example of de Sade demonstrates well the protean transgressive underpinnings of the erotic, as announced in his dicta, "crime is the soul of lust" (qtd. in Beauvoir 1966, 28). Because of the underlying principle of transgression, pornography is but one member of a family of symbolic transgressive activities. As noted earlier, the horror tradition has an erotic-transgressive impulse, and the violent and excessive bloodshed of the slasher genre constitutes both a superlative act of social transgression and, to move to the bond of signification that links the transgressive to the transcendent, an erotic experience in which the "the body overcomes its isolation and reestablishes itself as part of the greater world in a moment that is both orgasmic and transcendent" (M. T. Carroll 1993, 44; cf. Black 1991, 205, and Bataille 1957).

But popular modernity considerably problematizes the tangle of eroticism beyond the Bataillean critique and its largely pretechnological/ anthropological theoretic base. Whereas Bataille remarks that transgressive urges come from our condition as "rational being[s] who try to obey but who succumb to stirrings within [ourselves] which [we] cannot bring to heel" (1977, 40), in Fordist and post-Fordist culture we are dealing with the mass marketing of transgression, which ultimately means, paradoxically, obedience to transgression. Moreover, the transgression we speak of here is generally semiotic—textual or imagistic transgressions within the confines of Debord's spectacle and Baudrillard's simulacra. With this in mind, we must reject the oft-repeated idea that commercial culture reproduces the conditions of prostitution. A far better metaphor/metonym is found in pornography. As Haug explains, prostitution takes place at "the level of simple commodity production or rather a service industry, with the pimp as capitalist agent and the brothel as factory" and with use-value reckoned in terms of actual physical contact. In pornography, on the other hand, "the mere picture or sound, or a combination of both, can be recorded and reproduced on a mass basis, on a technically unlimited scale, which is restricted in practice only by the market" (1986, 55). In this reencoding of the economics of pleasure, satisfaction is bequeathed by an illusion, and thus may be called "illusory satisfaction"—which can result in "further demand [and] produce a compulsive fixation" (55). Thus, commercial culture simulates the conditions of pornography, in which appearance takes precedence over possession, passivity over activity. Rather than making simple and indirect obedience to authority the primary goal, it involves its subjects, paradoxically, in an obedience to transgression—an obedience that it does not reveal itself as such, for it shows, in the configuration of managed transgression, nothing of its authoritarian mechanism.

In capital's search for profits, ultimately, every site of transgressive potential must be exploited. Transgression even forces its way into children's commercial culture, which may be especially remarkable, given this genre's connection to moral didacticism since the publication of what is arguably the first work of children's popular literature, *The Renowned History of Little Goody Two Shoes* (1765), generally attributed to the novelist Oliver Goldsmith. But as Alison Lurie's work demonstrates, elements of subversion abound in both children's folk culture, particularly skip-rope rhymes, and children's commercial culture. This tendency is made more complex by MacDonald's observation that there is a tendency in popular culture to blur the lines between childhood and adulthood, resulting in what he calls "adultized children and infantile adults" (1957, 66), and thus we shouldn't be surprised to find a relationship between the reencoding of obscenity in the early 1970s and the appearance of more permissive forms of adolescent literature; Molz, for example, indicates the relationship between the Supreme Court overturning of a local ban of the 1971 film

Carnal Knowledge and a wave of these new works of children's literature (1978, 200–202). In the 1990s, a typical scenario for television commercials for chewing gum (the oral pleasure consumer product in the children's market that parallels the adult cigarette and is likewise eroticized) is the boredom of the classroom, which is countered by the pleasure and excitement brought by the forbidden object, gum (thus Starburst gum, with its promise of "bursts" of flavor, parallels the Newport cigarette, "alive" with flavor). Similarly, the mildly transgressive quality of R. L. Stine's "Goosebumps" book series have become bestsellers in the children's literary market (and the object of attempts to ban them). A more "hard-core" rendering of this form of children's culture is found in the "Barf-O-Rama" series (mid-1990s). If blood is the central fluid of the slasher film and semen of pornography, then in the "Barf-O-Rama" books (whose titles include *The Legend of Bigfart*, *Mucus Mansion*, and *Dog Doo Afternoon*), vomit is the iconographic center. Indeed, there is a strange similarity between the cum shot of pornography and the "vomit shots" in Pat Pollari's *The Great Puke-Off*:

> I imagine I looked pretty wild. There was gack dripping from my mouth. Allie, who is normally kind of pretty and has nice red hair, now had chunks of spew in the ends of her hair. (1996, 6)

But in spite of the ongoing phenomenon of marketed transgression, a fundamental economic paradox remains. That is, the titillation of pornography and pornographized popular culture is based on the "consciousness of transgression" that it engenders, and its commercial viability depends on that. As revenue is generated, however, markets expand and force their way into less restricted domains, thus causing the products to lose their transgressive value. As Michelson notes, "[T]he greater the general public disapproval, the greater the eroticizing potential," and thus the marketing of transgression can only lead to the destruction of our sense that a given product is transgressive (1993, 236).

Where does this leave us? Perhaps with Francis Fukuyama's thesis regarding the "end of history." That is, in capitalism's victory over alternative forms of social organization, problems of ideology are replaced by problems of management, and in a consumerist society the management task is largely that of keeping the cycle of consumption turning as transgressive spectacles are consumed, lose their erotic power, and are replaced by new spectacles, new imaginary transgressions.

5

Vocality, Visuality, Alterity: Black American Cultural Production

From what we've covered thus far, we can conclude, among other things, that cultural experience is often implicated in the Self/Other dialectic, and that cultural practices and genres that are often assumed to be at the margins may actually be at or near the center. These points lead inevitably to a consideration of the Other within the dominant-hegemonic structure of popular modernity. In the United States that means a consideration of black American culture, for no accounting of American popular culture, even a sketchy and selective one such as this, would be complete without a probing of the dialectic of race. In this chapter, we will start with a consideration of class and essentialism in critical discourse as a prelude to a discussion of black American cultural production in the era of popular modernity. We will be particularly concerned with two phenomenological orientations we've already examined in different contexts: that of voice and that of image; both acquire unique parameters when they mediate the American dialectic of race. Our focus genres here will be rap, a music/discourse emerging in the 1970s and employing a newly devised technological matrix, and popular film—in particular, films with a Hollywood narrative that is largely at odds with rap's informing ethos.

Race, Class, and Culture

In *Black Bourgeoisie*, E. Franklin Frazier argued that there are but two cultural traditions for the black American, "one being the genteel

139

tradition of the small group of mulattoes who assimilated the morals and manners of the slaveholding aristocracy; and the other, the culture of the black folk who gave us the world of the Spirituals" (1957, 98). This historical fact informs a mythohistoric belief—the belief that the black "folk" (proletariat, poor, underclass) are the guardians of cultural authenticity. The assumption seems to be that "middle-classness" and "cultural authenticity" are mutually exclusive categories, and that "authenticity" is a self-evident term. Not surprisingly, then, the ongoing discourse concerning black American culture has often centered on the problem of class and authenticity, and a continuum of positions has been generated.

At one end we have a number of related arguments for West African culture as the source of black American authenticity. These views receive much of their impetus from Melville Herskovits's *Myth of the Negro Past*, which argues that in spite of the presence of West African cultural variation there are a series of "underlying cultural similarities that support local variation" (1941, 55). In the 1930s and 1940s, Herskovits conducted a series of studies of African American cultures in Surinam, Brazil, and Haiti, and while he could not posit any broad general theory to account for the transmission of African culture to the New World, the evidence did suggest to him the survival of West African cultural traits, which he referred to as "elements." Mintz and Price largely agree; they argue that because black captives were drawn from a massive geographical area containing a great many cultures and languages, and because these peoples were "recombined" (a technique used to minimize the possibility of insurrection), it is difficult to maintain that there was a black African culture that had been transplanted to the New World "in the sense that European colonists in a particular colony can be said to have done so" (Mintz and Price 1976, 2). However, a degree of cultural continuity of the kind posited by Herskovits was clearly evident, and the specifics and the political implications of this cultural continuity have been probed by the "Afrocentric" school. The latter has its origins in the early twentieth century in the work of West African nationalist Edward Wilmot Blyden and manifests itself in late-twentieth-century scholarship in the work of Molefi Asante, R. F. Thompson, and others who argue that an African aesthetic informs a broad range of cultural practices throughout the diaspora.

At the other end of the continuum we have a series of viewpoints that privilege the creativogenic nature of slave society and the emergence of a new culture. Gunnar Myrdal, writing in 1944, concluded that the trauma of the Middle Passage kept African culture from leaving its imprint on the black American, who was, in his view "an exaggerated American" whose values are "pathological" variations on American values. Studies conducted by Stamp, Glazer, and Moynihan have been in agreement with this view (qtd. in Blauner 1970, 348). A similar view was later advanced by Ralph Ellison, who saw the real search for roots as leading to the American South rather than to Africa—a view that brought him considerable criticism from emerging Afrocentrists in the early 1970s (Ellison 1995, 235; cf. O'Meally 1980, 180). An interesting variation of this notion is found in Alain Locke's "The Legacy of the Ancestral Arts," in which he claimed that a "curious reversal" took place in the transplantation of the African to America: "[T]he characteristic African art expressions are rigid, controlled, disciplined, abstract, heavily conventionalized; those of the Afroamerican,— free, exuberant, emotional, sentimental and human" (1925, 254).

In most of these notions of West African and black American cultural authenticity, there is a injunction (sometimes explicit, sometimes implicit) in favor of "the black folk" rather than the black middle class of Frazier's "genteel tradition"; that is, if there is black American cultural authenticity, its origins are in the field, not in the house. But this is contestable, particularly in light of the most recent stage of this ongoing dialogue—the poststructuralist critique of essentialism, which promotes a skeptical stance toward the categories of "culture" and "authenticity." In this vein, Stuart Hall cautions that essentialism mistakes "what is historical and cultural for what is natural" and thus paradoxically valorizes "the very ground of the racism we are trying to deconstruct" (1992, 29–30). In like fashion, the historian Wilson J. Moses has exposed the ideological grounding of what he calls "sentimental afrocentrism" (1998, 18), while Cornel West asks, "[W]hat is black authenticity? Who is really black? First, blackness has no meaning outside of a system of race-conscious people and practices. . . . all people with black skin . . . have some interest in resisting racism," and yet, "how this 'interest' is defined and how individuals and communities are understood vary. Hence, any claim to black authenticity . . . is contingent on one's political definition of black interest. . . . In short, blackness is a political and ethical construct" (1993, 25–26).

In spite of all this uncertainty—or rather, because of it—one thing is certain: the assumptions articulated in Frazier's thesis open upon an epistemological uncertainty; questions concerning integrity and authenticity are, in black American cultural life, sites of anxiety regarding both race and class. White and White provide detailed historical examples of this in terms of clothing and body kinesics, noting that a class-based division has been evident in African American culture for some time. For example, they cite Hortense Powdermaker's observations of African American life in Indianola, Mississippi, which reveal that during the 1930s a division between middle-class and lower-class taste (with the "lower" taste marked by "bright" and "loud" colors) was in full force, and middle-class blacks "were keen to distance themselves from the dress and demeanor" of their lower-class counterparts, which, in their opinion and in that of middle-class whites, were characterized by vulgar and offensive "sartorial and kinesic excesses" (White and White 1998, 166, 222). This dilemma is the object of critical fiction, in, for example, the work of Alice Walker and Toni Morrison. In Walker's story "Everyday Use" (1973), cultural and class estrangement is not between, as in Wright's *Native Son* (1940), powerful ruling-class whites like the Daltons and impoverished blacks like the Thomases (two families who, as Mary Dalton points out, live in the same city and yet know absolutely nothing about one another); rather, the estrangement is generational—between a mother, steeped in the traditions of the rural South, and her college-educated daughter, who has taken up an Afrocentric pose, here rendered as a bourgeois rejection of what she, the daughter, perceives as a mundane (i.e., rural, Southern) and therefore undesirable cultural authenticity. A related conflict is found in Morrison's *Song of Solomon*, in which the protagonist is given the ultimate in bourgeois advice from his father: "[L]et me tell you right now the one important thing you'll ever need to know: Own Things. And let the things you own own other things. Then you'll own yourself and other people too" (1977, 55). In spite of his father's advice, however, he "couldn't get interested in money. No one had ever denied him any, so it had no exotic attraction. . . . He was bored" (108). Thus, the bourgeois values that have obsessed his father bequeath the protagonist a dull cultural sterility that he attempts to counter in his quest for an originary (and contrabourgeois) mythos. And indeed the protagonist's very name—Milkman Dead—seems to be a symbolic key to this cultural sterility.

The question of black class and authenticity has also surfaced repeatedly in recent nonfictional writing, and clearly, the ideological conflicts expressed by these texts are shaped by specific economic circumstances, such as the increasing chasm between the black underclass and the black middle class, particularly after 1970, and the fragility of the black middle class in its limited access to a variety of wealth-producing structures (Oliver and Shapiro 1995, 92). This situation is made all the more difficult by the continuing American pigmentocracy, in which assimilation efforts, no matter how successful, are halted, especially by what Lipsitz terms the "increased possessive investment in whiteness" that has been generated by the economic pressures of deindustrialization and that has fueled "a discourse that demonizes people of color . . . while hiding the privileges of whiteness" (1995, 379).

Ella Bell has termed the peculiar dilemma of the black middle class "bicultural stress"—the anxiety experienced during the assimilation process (1990, 459). Journalist Leantia McClain provides an eloquent exposition of this dilemma when she examines the accusation leveled at the black middle class to the effect that they have "sold out" and become "oreos" while whites are similarly ambivalent and often condescending to the black bourgeoisie. As McClain says, "[W]hites won't believe I remain culturally different; blacks won't believe I remain culturally the same" (1980, 21). Shelby Steele has also written on this situation when he talks of the "powerful double bind" faced by the black middle class, a bind between "the middle class values by which we were raised [that are], in themselves, raceless and even assimilationist" and the post-1960s imperative that "urges . . . an emphasis on ethnic consciousness over individualism" (1988, 43). Sam Fulwood echoes these sentiments. After stating that assimilation caused blacks to lose their "soul and rhythm," he concludes, "I stand wobbly, unaccepted by whites who do not regard me as their equal and hovering aloof from poor blacks, separated from them by a flimsy wrapper of social status. I straddle two worlds and consider neither home" (1991, 58).

These concerns, of course, are not only mediated through texts; they are also dealt with in various ways in the media that center on image and sound.

First, the image. George Lipsitz reminds us of the incident in which a French journalist asked Richard Wright about the "Negro problem" in America, whereupon Wright quipped that "there isn't a Negro problem; there is only a white problem" (1995, 369). Indeed, the image of the black American that has been generated by popular culture industry is largely a "white problem." These practices are as old as the nation itself: in two plays—*The Divorce* (1781) and *The Triumphs of Love* (1795)—citizens of the fledgling republic were introduced to Sambo, who provided comic relief through his nonsense songs and gaudy dress. Joseph Bodkin notes that Sambo was in fact "an insidious type of buffoon" illustrative of a "devious and encompassing" form of "social control . . . humor as a device of oppression" (qtd. in Dates and Barlow 1990a, 13). In the ensuing years, American popular culture generated that peculiar form known as blackface minstrelsy, which by the 1840s had evolved into a "formalized entertainment genre which imposed on the American psyche a very damaging perception of the black man" (Dates and Barlow 1990a, 7). Also during this period—a warm-up phase for popular modernity's matrix of technology, consumer culture, and political ideology—there is the example of Steven Foster, arguably the nation's (and popular modernity's) first songwriter, whose condescending portrayal of blacks contributed to this developing racism. An apt example of this is found in the now-forgotten second verse of "Oh, Susanna," with its astonishing genocidal image: "I jump'd aboard the telegraph / and trabbled down de ribber, / de lectrick fluid magnified, / and kill'd five hundred Nigga" (qtd. in K. Emerson 1997, 14). As Gilmore suggests, if the telegraph came to metaphorize the national body, then Foster's lyric can be seen as presenting an ideal of racial purity for that technological body. (This image of the victimization of the black body by "white" technology from a lyric from the 1840s is found again in a critical context in the 1940s, in Ellison's *Invisible Man* [1947], when the white doctors attempt to "correct" the black protagonist through electroshock.) In short, the development of the black image in American popular culture during the nineteenth century was (and continues to be) a "white problem" precisely because it has less to do with the representation of blacks than it has to do with the imagistic reification of the myth of white supremacy.

These practices continued into the film era. The one-reel comedies made between 1910 and 1915, for instance, featured the character "Rastus," who was really a Sambo for the age of film. Arguably, the imagistic reification of the mythos of white supremacy became more pronounced and damaging, given the broad audience for these films, given their use as a kind of school of American cultural literacy for newly arrived European immigrants, and given the very nature of the medium and the contemporary approach to its technical problems. That is, the movies added "a larger-than-life dimension to these stereotypes; films, short and without sound, depended on an exaggeration even greater than that practiced in the contemporary theater to make clear what was happening on screen. . . . this style of playing only heightened the ludicrous aspects of the black as comic stooge" (Leab 1975, 14). And while these exaggerated racist portrayals have largely been eliminated, the persistence of this image warfare is summarized by rapper Ice Cube in a 1993 interview with bell hooks: "It's hard to be black in America. Look at all the images that run across us, from television, school, just everything in general. . . . They put everybody in such a bad light. . . . It's mainly their fault, our self-hate . . . because all these images of white TV, that's the only thing we see. So when we look in the mirror, we changin' our hair, we changin' our eyes . . . try to not be black" (Ice Cube 1994, 127).

In the emergence of the feature-length film, the black image faced an even greater threat. The most technically proficient film of its day and a forerunner of Hollywood narrative style was D. W. Griffith's *Birth of a Nation* (1915), in which racist images (the story is taken from Thomas Dixon's white supremacist novel *The Clansmen* [1905]) were injected into a new and psychologically powerful medium. In response, the black middle class, led by an angry, galvanized NAACP, launched a campaign against the film via a coalition made up of "white, Jewish, urban, philanthropic, socially conscious elite[s]" (Cripps 1990, 135). They forced Griffith to alter certain aspects of his production, and an ad hoc committee of powerful black notables, most eminently Booker T. Washington and his "Tuskegee Machine," responded with their own epic, *Birth of a Race* (1918). In the ensuing decade, the response took two forms: first, the lobbying efforts of the NAACP and the criticism of the black press; second, the efforts of other segments of the black middle class to control the process through the creation of their own production companies. Significantly, one of the most successful efforts, *The Scar of Shame* (1927), produced by the Colored Players Firm of Philadelphia, set class against class. As Cripps explains, the middle-class blacks in search of the finer things in life were given heroic casting in comparison to the "low down" Negroes (1990, 144). It is worth noting that in a recent essay, Amiri Baraka remarks that he regards Spike Lee as part of this "school": "Like Ellison and Ishmael Reed before him, they feel that only the black middle class, including the 'crafty' house slave[,] is *dignified*. We can, as Mao said, identify in art a class stand, an audience (for whom?), and also those whom the artist implicitly praises and condemns . . ." (1993, 147). But we get ahead of ourselves: the example of Oscar Micheaux notwithstanding, it would be a long time before there was anything like a black American cinema. The destruction of the European film industry during World War I put Hollywood in a mode of capitalist expansion, and inevitably its product was, as Cripps points out, "corporate, studio-dominated, risk free, vertically integrated," and roles for blacks were thus required to confirm, for the most part, "the enduring system and the place of white people in it . . ." (1990, 137).

Essentialism, Economics, and Music

Black popular music proceeds in a manner that parallels the evolution of the black image, though there are significant differences. First, there is the primacy of music in African American cultural history. While a common language and the means to produce in any of the plastic arts was drastically limited, the ability to create music, particularly vocal music, remained. And even though drums were "outlawed" by plantation owners who knew of the African "talking drum" technique and feared that it might be used organize an interplantation insurrection, blacks responded by developing other rhythmic techniques—hand clapping, spoon playing, and so on. As a result of these factors, music has been a major element in the mythos of black cultural authenticity; Du Bois, in his discussion of spirituals, commented that they are "the siftings of centuries; the music is far more ancient than the words" (1903, 254), thus proposing that under an outer skin of Western culture (Christianity, the English language, and a more or less Western musical format) there is an essential West African core. This isn't as nonproblematic as it may seem at first, for here Du Bois invokes the long-standing concept of music essentialism, which has particular ties to protofascist currents in the late nineteenth century in the heady combination of romanticism, nationalism, mysticism, and music. A benign example is Dvořák; more virulent strains are found in Wagner's project of constructing a German national mythos informed by medieval legend and, to some extent, the racial theories of his contemporary, Arthur Gobineau, whose work in the late 1850s formed the basis of modern racism. And thus we find in Wagner's thought an evolving fascist imperative of demonizing the Other, constructed with a view toward generic purity—

a true German music that would perforce exclude the "degenerate" Jewish influence, particularly that of Mendelssohn. The difference, then, between Wagner and Du Bois is the difference between the essentialism of an aggressive dominant and that of the assertive subaltern.

The assumptions of a cultural critic like Stuart Hall are very different from those of Du Bois, and yet both share the notion of the primacy of music in black diasporic culture. In Hall's analysis, the black subject was largely outside of the logocentric ideology in which the "direct mastery of cultural modes meant the mastery of writing," and thus, in opposition, the diasporic subject "found the deep form, the deep structure of their cultural life in music" (1992, 27).

Because music is a nonvisual medium, the politics of racism, at root a visual matter, operates somewhat differently than it does in film. Michelle Wallace sorts out this problematic, noting that the black image is a paradoxical combination of visibility ("the very markers that reveal you to the rest of the world as visible") and the phenomenon of "invisibility," defined by Ellison in *Invisible Man* as "a matter of the construction of the *inner* eye, those eyes with which they look through their physical eyes upon reality" (1947, 3). This paradox, according to Wallace, informs the fact that a black tradition in visual culture as "compelling to other Americans the Afro-American tradition in music" has not been forthcoming (1990, 40). Whereas in the early history of the black visual image, white minstrels covered their faces with burnt cork in order to appear black, black music had undergone an opposite operation in order to allow black music, or a somewhat diluted and modified version thereof, to have a white face; this is what Garafalo calls the ongoing phenomena of "black innovation/ white popularization" (1990, 104), a subject we touched upon in chapter 2 in our discussion of Bing Crosby and the assimilation of jazz into pop vocal style. And in the jazz era, the great black bands were overshadowed by the sounds of Paul Whiteman (indeed, a white man); the rhythm and blues (a term that replaced the increasingly unacceptable "race music") of Ruth Brown and Little Richard was commercially eclipsed by Elvis Presley in the 1950s and by the Beatles in the 1960s; and in the 1970s, disco, a black and Latino music, was represented to the mainstream by a white Australian band, the Bee Gees (105). As Johnny Otis tartly put it, blacks create musical forms only to "get ripped off, and the glory and the money goes to white artists. This pressure is constantly on them to find something that whitey can't rip off" (qtd. in Garafalo 1990, 80). And as Rose notes, this "whitewashing" continues in some unexpected places, such as the use of the term "dance music" in contemporary music criticism (1994, 84). In spite of this, there has on the whole been greater representation of blacks in sound (rather than in image), perhaps because a disembodied black voice is more tolerable in a racist society than a black image. Thus a peculiar daily occurrence during the days of legally sanctioned apartheid in America was the presence of (disembodied) black recorded music in places where (embodied) black people were not permitted. Compare MTV's resistance in the early 1980s to music videos by the eminently marketable Michael Jackson, who at the time had the best-selling album in history (cf. Q. Jones 1995).

The emergence of hip-hop culture—rap and go-go music, break dancing, graffiti art, basketball as an iconically central sport (cf. Boyd 1997, 112–27, esp. 119), new slang and clothing styles—in the mid-1970s is complex and overdetermined. In many respects it is traditional: in rap, the persistence of vocal invention and a *bricoleur* approach (i.e., assembling an aesthetic inventively with the materials at hand) to percussion (the "scratching" technique are the latest in a long line of African American percussion innovations) suggest that continuance. As David Toop notes, the new form of musical expression that appeared around 1975 in the streets and nightclubs of the Bronx had immediate connections with disco, funk, the radio DJs, The Last Poets, Gil Scott-Heron, but even further back than that, to "the bebop singers, Cab Calloway, Pigmeat Markham, the tap dancers and comics . . . ring games, skip-rope rhymes, prison and army songs, toasts, signifying and the dozens, all the way to the griots of Nigeria and Gambia" (1984, 19).

And yet, rap is more than a continuation of older forms. Indeed, Tricia Rose warns us that if we think of rap only in this way, we "romanticize and decontextualize [it] as a cultural form" (1994, 95). Rose here refers to two interrelated contexts: first and fundamentally, the social impact of de-industrialization; second, the possibilities provided by new music production technologies.

In the early 1970s, a number of socioeconomic factors, alluded to previously with reference to the Oliver and Shapiro study, provided the contexts for the development of hip-hop. De-industrialization is the shorthand term for the economic disruptions that began at about the midpoint of the Vietnam War, and may be summarized as follows:

first, the economic disruption wreaked by the cost of the war itself; second, the evaporation of working-class and lower-middle-class manufacturing jobs (entailing the erosion of the urban industrial sector) as that sector of the economy was increasingly relegated to foreign markets and as a postindustrial "service" economy began its ascent; third and relatedly, the rise of the Japanese automobile industry; fourth, the energy crisis and the Arab oil embargo, which served to increase the popularity of Japanese imports while the auto industry's decision makers in Detroit, living under the American myth of industrial omnipotence, continued to manufacture gas guzzlers. These disruptions had additional specific impacts on the urban black community; in particular, the evaporation of federal and city funds for inner-city social services and education (including traditional music education), the development and gentrification of sectors that had formerly been marked for low-cost housing, the destruction of communities (particularly, as Rose demonstrates, in the case of the Bronx freeway project). In Rose's view, these social ruptures and dislocations form a structural homology with the basic aesthetics of hip-hop (in which flow, layering, and rupture are prioritized in sound, music, and the images of graffiti art); these aesthetics "suggest affirmative ways in which social dislocation and rupture can be managed and perhaps contested in the social arena" (1994, 39).

The de-industrialization phase and the emergence of rap takes us back to our discussion of the black middle class. As noted previously, in the commentaries of Steele, McClain, and Fulwood we find evidence of an intensification of the black American class schism. The de-industrialization phase that has exacerbated the this schism, and the rise of hip-hop culture illustrates this, for the black middle class seemed as mystified and alienated by this movement as was middle-class culture in general. Black radio totally missed the rap breakthrough year of 1989 (Barlow1990b, 244), and so did the print media; as rapper M. C. Shan put it, "[I]f a white rock 'n' roll magazine like *Rolling Stone* . . . can put a rapper on the cover and *Ebony* and *Jet* won't, that means there's really something wrong" (qtd. in Dyson 1993, 7). The "wrong" that Shan refers to is well summarized by Ernest Allen: "[O]ut of the turmoil of poverty, drugs, and death emerged an African American youth weltanschauung at considerable odds with those of the more stable black working and middle classes" (1996, 163). This schism is sometimes reflected as a theme, or a metatheme, within rap, as in Ice

Cube's "Turn Off the Radio" from the album *AmeriKKKa's Most Wanted* (1990). The track begins with a sample (from Spike Lee's *Do The Right Thing* [1989]) of a white bigot delivering a stream of racist epithets, which is immediately followed by the smooth voice of a black R 'n' B disk jockey promising that rap won't be heard on this R 'n' B-only station. Later in the song, we hear the stereotypically white voice of a radio listener talking about how he hates rap . The song ends with the pleasant voice of a news announcer telling us about the mortality rates of young black men.

The second context for hip-hop that Rose articulates is techno-logical. The primary technology of rap—"sampling," first introduced by Afrika Bambaataa and other DJs in the early days (early- to mid-1970s) of Bronx-style rap—involves taking segments from various musical recordings and using them as the raw material for a new sound. This was first accomplished primarily through the use of two turn-tables to artificially extend the "break beat" on a phonodisc; as rappers entered the studio and new technologies evolved (especially digital sampling machines) this became an increasingly complex method of organizing a densely intertextual discursive soundscape. Rap producer Hank Shocklee's conceptualization of this process is markedly similar to European avant-garde concepts, such as Edgar Varèse's "music concrète" (cf. Rose 1994, 80–81). Richard Schusterman, in an essay canvassing the postmodern aesthetics of rap, sees in the techniques of rap (cutting, mixing, scratching) "a variety of [postmodern] appropriation, which seem as versatilely applicable and imaginative as those of high are—exemplified by Duchamp's mustache on the Mona Lisa, Rauschenberg's erasure of a De Kooning canvas, and Andy Warhol's multiple re-rep-resentations of prepackaged commercial images" (1991, 617). In its musical form and production processes, rap, then, is traditional, tech-nological (significant, given the early examples of the technocracy's exclusion of the black body, as mentioned earlier), and, as Houston Baker points out, "a deft sounding of postmodernism" (1993, 89).

But digitally reproduced instrumental music is only one part of rap; it is the voice of rap that is perhaps more important, for the foregrounded nature of the speaking voice in this form is perhaps what led its critics to insist that it "is not music," perhaps implying that, if anything, it is a mode of discourse. The fusion of voice, discourse, and technology that comprises part of the rap aesthetic is evidenced by the fetishization of the microphone. As with Bing Crosby in the 1930s, it is common for rappers to focus on microphone technique, as reflected in the overt theme of a number of rap lyrics (Rose 1994, 55). In our previous discussion of the human voice in terms of the rise of commercial radio and the subsequent rise of authoritarian politics, we noted Cantril and Allport's contention that the disembodied voice of radio creates a particularly powerful experience for the listener. The voice of rap, like the voices of the fascist leaders of the thirties and some of the talk show radio stars of the 1980s and 1990s, is a voice of authority, and this is the focal element in rap's ability, primarily through the vocal medium, to provide its listeners with a strong experience of identity (cf. Frith 1991). The nexus of performance, listener, and the construction of personal identity here are linked to the larger construction of mythohistoric national identity.

We will recall here the observations of Hobsbawm and Anderson to the effect that modern nationalism is largely "invented" through the organs of public education, public monuments, and public celebrations. In the United States, popular culture has been for the most part in complicity with the invention of the nation. A sterling example is that of Horatio Alger, whose popular novels (e.g., *Ragged Dick* [1867]) promoted an ideology that was both Christian and capitalist as well as providing a template of behavior, a model of citizenship for industrial America. In the postindustrial state of anomie in which the promises of Alger's ideology seem increasingly dated and empty, marginal cultural practices like metal and hip-hop in their formative years posit quasi counternationalisms to fill the vacuum.

Rap strongly reflects this encoding of counternational identity. Ever since rap's crossover breakthrough—Grandmaster Flash's "The Message" (1982), a jeremiad with a strong voice telling about life in de-industrialized urban America from a decidedly emplaced view (i.e., one oriented toward local neighborhood affiliations)—the genre has been dominated by strong and authoritative styles, as in the neonationalist raps of Chuck D. of Public Enemy ("Fight the Power"), and the gangsta (and "gangstanationalist," to use Kelly's term) raps of Ice T, Dr. Dre, and Ice Cube. One might also note the matriarchal authority of Queen Latifah, the feminine/aggressive modality of the Lady of Rage, and the attempts to outdo male sexual bravura in MC Lyte, Yo-Yo, TLC, and Foxy Brown. We see that while there is a wide range of ethical and rhetorical stances, they are all informed by a centered and authoritative position as conveyed through vocal dexterity and timbre.

Ice Cube has been particularly adroit at using a powerful and authoritative voice to articulate ghettocentric discourses comprised of both an ethos of antisocial swagger and political awareness. We should note, however, that his reality-based attack on "the media's complicity in equating black youth with criminals" is itself complicit with ideologically based expressions of masculinity and sexuality that "constrain [the effort] to create a counternarrative of life in the city" (Kelly 1994, 185). His album *AmeriKKKa's Most Wanted* begins with the sound of a death-row inmate being led to his execution. After the inmate contemptuously and confidently delivers his last words ("fuck all y'all"), the switch is pulled and we hear a surge of voltage. The voice continues—as though defying death—into "The Nigga Ya Love to Hate," a song that is all at once a call for political awareness, an attack on the various factions who have an interest in suppressing true "street" knowledge (the police, the media, and complacent black entertainers), and a manifesto. Its black male braggadocio invokes an entire litany of characters from black folk narrative—Dolomite, Shine, the Signifying Monkey, Stack o' Lee—the whole "Bad Nigger" mythos first identified by Richard Wright in *Native Son*. Another good example of this is found in Ice Cube's duet with Dr. Dre (on the 1996 *Murder Was the Case* album by Dre protégé Snoop Doggy Dog), "Natural Born Killaz," in which the two rappers—in a polygeneric move that, like a number of rap songs, combines the transgressivity of urban crime with that of the horror genre—bring us on a journey into the mind of a cold-blooded killer who is driven by sadistic urges. By centering on the speaking voice in a way that differs from traditional "sung" music, rap lays claim to ethical and aesthetic properties that, as we shall see, do not easily transfer from their particular media/generic constraints. But before moving to a consideration of the way rap has been represented in the Hollywood film, we need to briefly consider rap's "white" counterpart, heavy metal, and how these two musics reflect the dialectic of race, class, authenticity, and essentialism.

Rap and Heavy Metal: Parallel Discourses of Subalternity

Essentialism's construction is essentially binary (in this case, "white" and "black") and symbiotic, for a postulated Self knows itself only through a postulated Other, as we noted in our discussion of the aesthetics and phenomenology of fascism. For this reason, it would be useful to consider the white essentialism of heavy metal music, a subgenre that is different from rap (particularly in terms of its initial target audience) and yet in many ways similar in that its popularity is rooted not only in essentialism but in de-industrialization and counternationalism.

Although there have been a number of attempts to cross-pollinate the rap and heavy metal genres (as in Living Color and Ice-T's metal band, Body Count), they are generally thought of as being essentially different. If this is not entirely true today (with rap-rock bands like Rage Against the Machine), it was most certainly the case during their formation in the early 1970s. If we bear down on this uninterrogated assumption, we see that behind this category of "essentially different" lurks racial *essentialism*, for while metal is performed by and for whites, rap is for the most part a black performance art. (Ice Cube, for example, proclaims that his music is for blacks, and that whites who listen to it are "basically eavesdropping" [1994, 129].)

American popular music with a black, Southern base became a site for a reenactment of the late-nineteenth-century discourse regarding music and essentialism. White racist critics of R 'n' B wanted to stop the influence of "nigger music" on white youth. Less obviously racist critiques were leveled from other sectors of the culture: both Arnoldesque and Adornoesqe critics attacked black-based music as being infantile, inferior, mindless, primitive, and generally corrupting.

In the early 1980s when rap was gaining in popularity, rock musicians, proud of their musical abilities, were prone to criticize rap for its lack of traditional musicianship (cf. Rose 1994, 81); among some music fans and musicians, "rap is crap" became a common slogan, one that rock musician Greg Allman one repeated (Allman 1995). (This attitude has persisted through 1998, as evidenced by the "Metal Up Your Ass: The Why Rap Sucks Page" on the Internet.) In this instance, we have the strange phenomenon of white practitioners of one branch of popular music with African American roots—rock—rejecting the contributions of an emergent black form, thus displaying a curious historical amnesia and leading to the "black innovation/white popularization" model we noted earlier. The resistance to rap as it just began its "crossover" in the early 1980s formed an analogue to the canon debates in academe that were beginning to flare up at about the same time. Paul Lauter, a prominent spokesperson for the movement to reform the canon of American literature, noted that the canon of high modernist poetry, through it's adherence to aesthetic rules and concepts, excluded poets of the Harlem Renaissance. The phenomenon Lauter indicates here is not very different from the antirap contingent's use of "musicianship" as an exclusionary code—making them a kind of gatekeeper for the house of American popular music.

The struggle of American popular music is one in which a deep-seated ideology of apartheid and white privilege comes up against the contrary project of employing African American–based cultural expression in the creation of a postpuritanical, youth-oriented consumer culture. The "British Invasion" of the early sixties is an example of this, for it constituted a kind of historical amnesia regarding rock's origins in black Southern music. Heavy metal, which followed the British pop invasion after an intermediary period of psychedelia, certainly continued this. As Robert Walser explains, this "forgetting" can be found in the heavy metal mythos in rock journalism, which fails to trace African American influences by beginning only at "the point of white dominance." But, as Walser points out, "to emphasize Black Sabbath's contribution of occult concerns to rock is to forget" Robert Johnson and Howlin' Wolf; to "trace heavy metal vocal style to Robert Plant is to forget" James Brown; and finally, "to deify white rock guitarists like Eric Clapton or Jimmy Page is to forget the black American musicians they were trying to copy." In short, "the debt of heavy metal to African American music making has vanished from most accounts of the genre, just as black history has been suppressed in every other field" (1993, 8–9). The function of rock journalism of this kind, then, has been to reinscribe American music, and metal in particular, according to the code of "whiteness."

Heavy metal in the early 1970s appealed to a predominantly male, white, working-class and lower-middle-class audience who were entering the job market just as de-industrialization was eliminating entry-level manufacturing jobs. As Walser notes, the class origins of metal became obscured by its growing transclass popularity and commercialization, but the origins of the seminal performers and audiences are clear. For example, Ozzy Osborne and the members of Judas Priest are from Birmingham's industrial working class. The division between metal and mainstream rock is drawn along class lines, or at least perceived class lines. This is also true of the initial relationship between rap and

R 'n' B, and is evidence of a particular tension in the black community during the de-industrialization phase.

If the postulation of an economics of de-industrialization is valid, then we should be able to trace the structural homologies of economics on the specifics of generic formula for both forms of music. And indeed, it is not difficult to discern these parallels. In spite of their apparent differences, rap and heavy metal are structurally homologous in a number of obvious ways.

First, the lyrics of both these genres are aimed at the needs of disenfranchised youth. They both reject bourgeois parental values and frequently thematize this rejection; this is a feature of rock music generally, but in the case of these two genres, this rejection is more extreme. Second, they both speak the language of masculine sexual aggression. While the products of the culture industry can certainly be expected to reflect the sexism and patriarchy of the dominant, in both rap and heavy metal we find particularly extreme (and to many, offensive) expressions of such (which makes us wonder just how far from the mainstream they are). This aggressiveness toward women is part of a generalized backlash against feminism and the entry of women into what had been exclusively male sectors of the workplace. Third, they both value performativity and technological mastery, as demonstrated by the rap virtuosi (Fu Snickens, Busta Rhymes), guitar heroes (Richie Blackmore, Eddie Van Halen, Yngwie Malmsteen) and the ongoing refinement of electronic and digital techniques (in metal, feedback loops and various modification devices; in rap, the refinement of sampling techniques). Relatedly, they both prioritize loudness as a way of occupying aural territory and proclaiming identity (cf. Frith 1996). Fourth, both forms thematize death and destruction. Ice Cube's *AmeriKKKa's Most Wanted*, as noted, begins with the sound of a death-row inmate being led to his execution, and similar death enactments are found in Snoop Doggy-Dog, Tupac Shakur, and Dr. Dre. In metal, suicide is a frequent theme (one particularly identified with Ozzy Ozborne) and an issue that received public notoriety when the argument was made that Judas Priest had used "backward masking" to advocate suicide in their *Stained Class* (1978) album. Fifth, both forms enact a kind of quasi nationalism and fascism, with their attendant concepts of collective energy and retribalization. This strikes me as the most important link, for it is here that economics and essentialism

meet; economic dislocation can reawaken both black and white nationalism—the hip-hop nation of rap; the headbanger nation of heavy
metal.

One song that fully captures the kind of underground nation theme
with racial overtones is "One in a Million," from Guns 'N' Roses
G'n'R Lies (1986). The song's narrator is a white male teenager, a
runaway lost in the multicultural stew of Los Angeles. Through offensive slurs, he expresses contempt for the immigrants, blacks, and gays,
and seems to fear that these groups will either take over the country or
destroy it by spreading disease. The experience of Axyl Rose's small-
town white teenager describes well what George Lipsitz means by
"the possessive investment in whiteness," and much of white nationalism in the United States is driven by the perception among marginalized whites that they are not appropriately benefiting from the
code of white privilege as other groups find some degree of economic
success.

From the Bronx to Hollywood

A second black aesthetic movement that comes into existence shortly
after hip-hop is found in the black filmmakers' movement. In the
early 1970s a black independent film movement was established, centered at UCLA; a good example of their output would be Melvin Van
Peebles's *Sweet Sweetback's Badasssss Song* (1971) (cf. Bambara 1993,
118–19). During the 1970s there also emerged a broad commercial
market for so-called blaxploitation films, such as Gordon Parks Jr.'s
Superfly (1972); these films invariably featured a black hero "in opposition to a stereotypical white menace . . . bent on destroying the
African American community" (Boyd 1997, 89). Following this phase,
in the 1980s and early 1990s a number of new black directors and
producers came to the fore, including Mario Van Peebles, Doug
McHenry, Warrington and Reginald Hudlin, Spike Lee, Alan and
Albert Hughes, and John Singleton. Interestingly, these filmmakers,
like the more overtly political black directors of the early 1970s, were
for the most part (but by no means all; cf. Boyd 1997, 99) academi-
cally based. If rap, at least in its first wave from the Bronx, came from
the streets, it seems that black film came, as it were, from the quad,
which perhaps explains the considerable number of black films during

this period that center on collegiate life—Warrington Hudlin's documentary *Black at Yale* (1974), Spike Lee's *School Daze* (1988), George Jackson and Doug McHenry's *House Party II* (1990), and John Singleton's *Higher Learning* (1995). But another important influence on this school is hip-hop, which "has helped fuel the African American cinema resurgence in Hollywood" (Perkins 1996a, 1).

As hip-hop crossed over into mainstream popular culture, the interactions between rap and film became increasingly evident. The earliest rap films were quasi-documentary in nature, such as *Beat Street* (1984) and *Krush Groove* (1985)—with *The Show* (1996) continuing the "rapumentary" format into the Los Angeles "G Funk Era" of rap. A later set of films are constructed such that, while rap plays no real significant role diagetically, it does play a nondiagetic role as a soundtrack—the music, as it were, of criminality (*Colors* [1988], *Deep Cover* [1992]). A related set of films, informed by a social realist ethos (e.g., *South Central* [1992], *Boyz in the Hood* [1990]) used rap in a similar nondiagetic manner. Accompanying this was the increasing use of rappers in lead rolls, beginning with *Tougher Than Leather* (1987) with Run-D.M.C., the second-generation rap group from the black suburb of Hollis, Queens, who initiated rap's crossover to the mainstream (cf. Nelson and Gonzales 1991). A number of other rappers eventually embarked on film careers with varying degrees of success. Some played rap-oriented roles (Ice T, Ice Cube, Kid 'N Play, Treach from Naughty By Nature, Eric B., Rakim), while others made the transition to television and film roles having no particular relationship to rap (Tupac Shakur, Queen Latifah, Will Smith [who starred in *Independence Day*, discussed previously]). When they served in lead roles in rap films or those purporting to portray ghetto life, the presence of the rap stars promoted the films' claims to authenticity while guaranteeing the attendance of the already established audience for the music.

The transference of narrative material from a sound-based to an image-based medium is a complex process, largely because the Hollywood feature-length film comes with its own aesthetic priorities, which would prove resistant to the hip-hop aesthetic as articulated by Rose. The MTV video format has, because of its close weave with music that often dissolves the diagetic/nondiagetic boundary, more potential for transferring a hip-hop aesthetic into a visual medium. Consider the style of many rap videos, in which the rapper directly addresses the camera while using the body language and gesticulations of hip-hop and at the same time, quickly shifting his or her distance from the camera. A speaker-viewer spatial relationship is thus created that is not at all in keeping with the nonintrusive aesthetic of Hollywood style. A good example of this is found in Naughty By Nature's *O.P.P.* video (directed by Rodd Houston and Marcus Raboy), which compounds this mode of direct address with a neighborhood location—the "ghetto specificity of East Orange, New Jersey" (Rose 1994, 11). The use of a hip-hop aesthetic in feature-length films provides greater difficulties. The basketball sequences in *Above The Rim* (1994) would qualify, for they articulate the relationship between the sport of basketball (as reinvigorated and reencoded by Michael Jordan) and rap music (e.g., "slammin'" is a term used to describe both the physical events of basketball and the "aural event" of rap). Furthermore, as Victoria Johnson argues, Spike Lee's *Do the Right Thing* (1989) likewise effects a breakdown of the diagetic/nondiagetic polarity and employs a more immediate camera style, particularly in street scenes. It is also worth noting how well Lee's film symbolizes the dilemma of the black American in terms of the phenomenal categories of space, time, and image. For example, Sal's Pizzeria is a space where blacks may spend and consume, but they are not to be heard, nor are they to be reflected in the official pantheon of icons (the riot is sparked by Sal's violent destruction of Radio Rahim's ghetto blaster and his refusal to allow a few of the heroes of black American mythohistory to be posted on his Italian-American "Wall of Fame"). Another good example is discussed by Todd Boyd—the scene in which one of the neighborhood's few white residents accidentally steps on one of Buggin' Out's Air Jordans. An argument ensues in which an image-commodity representing blackness in contemporary American culture (i.e., Air Jordans = Michael Jordan) becomes the locus for a struggle involving space, image, and ownership. As Boyd observes, the scene illustrates " the disturbing reality . . . that Black people's ownership is often limited to trivial items that have no real value outside of demonstrating their slave-like relationship to commodity culture" (1997, 108).

What happens, then, in the shift from rap's oral narrativity to a visual one? As noted in our discussion of Ice Cube, the mythic and heroic stance of the rapper is a key element in the audience's identification with his or her voice and its willingness to let this voice occupy its phenomenal space. It is a voice that, as Perkins observes, summons up "romantic notions of ghetto authenticity" defined by unrelenting oppositionality. It has—and this is the key to rap's entry (and gangsta rap in particular) into commercial culture—become a source of identity for white youth who "search for a new identity in a post-Soviet, post-industrial, new world order. . . . As long as youth culture is dominated by the cult of the commodity, there will be a desire for the 'real'" (1996a, 20). Further, the preoccupation with authenticity is also a characteristic of white heavy metal culture, suggesting that such preoccupations have both class and racial determinants.

While it is difficult to assign ideological values or degrees of certifiable "authenticity" to media and genres in any simple analogical manner, it is also true that narratives and ethical and aesthetic codes cannot simply be translated from one set of relations-of-production to another without mutation. The material conditions of cinematic production/reception privileges narration from the outside, and we have become accustomed to this, the official eye of film narrative; but rap has a lyrical inner voice (inner voice as opposed to outer eye), constructed as emanating from the purportedly authentic subject. Hollywood cinema, then, is not an empty vessel into which the content of rap can simply be poured. Or rather, if it is a vessel, the contents poured into it will assume the vessel's shape, and thereby betray Hollywood's structural intent.

One factor that governs the transformation of the conventions of the rap mask lyric narrative into those of the Hollywood film (both allegedly having the same reality-reference, i.e., black urban experience) has to do with full-length narrative film's greater chronological scope and its ability to map out a life, qualities that it received from its prior informing discourse, the novel. The novel, an example of what Frye calls a "radically continuous form" (1957, 303), derives its character from the cultural and economic matrix of eighteenth- and nineteenth-century Europe. Calvino accurately identifies the novel as the narrative form for that period separating premodernity from postmodernity: the novel, he says, was a product of "that period when time no longer seemed stopped and did not yet seem to have exploded, a period that lasted no more than a hundred years" (qtd. in Harvey 1989, 291). The novel's informing base is capitalism and the bourgeois systems (e.g., established Protestantism) informed by capitalism, and thus the novel's narrative shape and ethical stance, particularly that of the *Bildungsroman*, typically stress self-transformation, socialization, economic success, and accommodation to the dominant. This

narrative code, with of course some modification due to media constraints, served as the model for the full-length Hollywood-style film;
this code can and has been challenged, but the weight of this tradition,
the economics of Hollywood production, and our tendency to con-
ceptualize film in these terms are powerful conservative forces.

There is, then, in Hollywood narrative a pull that is largely counter
to the rap narrative. Or one could say with bell hooks that the question of authenticity regarding rap is largely nonsensical, because once
rap is part of a complex, abstract capital system, it is largely denuded of
its marginal emplacement and has no real locus other than the market.
Rap-based film can be viewed a continuation of that displacement
(hooks 1996a). If we consider this point in the broadest possible manner, ideological struggle, since the beginning of the capital-techno-
logical epoch, have been mired in the struggle between *place* and *space*—
between, on the one hand, intimate places or regions that emerge
from experience, perception, and corporeal emplacement; and, on the
other hand, a more abstract space (as was discussed in chapter 1). The
nature of commodification is such that even though the introduction
of a marginal and "emplaced" (i.e., "niggacentric" and "ghettocentric")
discourse into the abstract spatiality of the capitalist market raises the
possibility of synchronous rebellions (as those that occurred in various
places in 1848 and 1968), the very appropriation of a more commanding space entails a displacement that ensures an impotent insurrection,
for oppositionality and insurrection have through this process become
commodities. This is really a variant of the process the phenomenon
of marketed transgression, as was discussed in chapter four.

A good place to witness some of these dynamics is in Nelson
George's account of the creation of the rap parody film, *CB4* (1993).
George, a black music critic who later turned his attention to the film
business, candidly exposes the restraints faced by the would-be producer. In 1984, George helped organize the Stop the Violence Movement, an ad hoc collection of members of the hip-hop community
and other concerned parties who wanted to critique rap's destructive
potential and try to turn it around. It occurred to George that humor
might be the antidote and that rap had come to the point in its development where it could be productively parodied. George began with
the central "scandal" of rap: "It was well known that many of rap's so-
called street reporters were actually middle-class kids whose records
were as fueled by blaxploitation flicks as real life." On the other hand,

George was concerned that "none of the rap-based films to date had captured" this music which celebrated "black existence, something the calculated pop iconography of Michael Jackson and Lionel Richie has sacrificed for an often dubious universality" (1994, 137). Clearly then, George entered the *CB4* project with an awareness of the authentic/inauthentic dichotomy within rap, and he hoped to address it critically and parodically. He found, however, that it is very difficult to hold onto one's intentions in the film business. Once Hollywood is involved, he says, "the stakes are raised, and no matter what good intentions people go in with. . . the project changes. You are guaranteed to see one fourth to one third of the original ideal sacrificed to the great God of commerciality" (138). Whatever oppositional essence he and cowriter Chris Rock may have wished to preserve from the world of rap, the film would ultimately be a Hollywood product: "[Y]ou can't be a real rebel in the black nationalists' sense, or be completely oppositional within the Hollywood system of compromise" (George 1994, 37; cf. hooks 1996b, 52–59 for an analysis of the appropriation of black film in the case of *Waiting to Exhale* [1995]).

George and Rock had originally planned the film around the concept that the fictional rap band CB4 (cell block number four), a group of middle-class youth who portrayed themselves as hardened street thugs and ex-convicts, would face a "comeuppance" when some probing journalists discovered that their claims were fraudulent and went on to excoriate them in the popular press. It is at this juncture that an interesting reversal takes place in terms of narrative construction and its ethical and ideological parameters. As George notes, he was persuaded by producer Brian Grazer (who has produced many popular films, including *Apollo 13*) that "some nasty magazine articles didn't constitute enough jeopardy to drive a movie, even a farce" and in response, George created *CB4*'s villain, an ex-convict and thug named "Gusto," based on someone George had known who was murdered in 1991 (George 1994, 143). As a result, the basic narrative concept for *CB4* became one in which three black middle class teenagers pass themselves off as hardened thugs from "the 'hood" in order to assure their success as "authentic" rappers, a scam that gets them in trouble with real ghetto gangsters.

In *CB4*, Gusto is the remorseless and irredeemable Other (not unlike the aliens in *Independence Day*, which indicates the deeply structured heterology of American popular modernity and its informing mythologies). Strangely enough, whereas in the rap lyric the criminalized antisocial subject is the precise target of the auditor's identification, in *CB4* this subject is "Otherized" (becomes the "Other's Other") and another character becomes the point of entry and identity for the viewer—an essentially middle-class character (one of the band members). The latter's transformation, in true *Bildungsroman* fashion, is one in which he sheds his street pose and affirms his middle-class identity in response to the probing of his father, who chides his son for being a poseur: "You're not from the street—I'm from the street. And only someone who wasn't from the street would glorify it."

The film begins with the rappers telling their story to a white reporter, and in film's conclusion, Gusto is handed over to the white policemen. Thus, the entire narrative is framed by the voice and authority of white institutions, which is itself a structural emblem of rap's transformation and perhaps, intentionally or otherwise, a comment on the racial problematic of this media/generic transformation. A second rap parody that came out at the same time is *Fear of a Black Hat* (1992). In this film, the narratee (the auditor within the story) is not a white reporter, but a light-skinned, academic, middle-class graduate student writing her thesis on rap. This parody, like *CB4*, has a narrative structure that includes a criminalized Other, an essentially middle-class central character, and whites or assimilated characters with an official institutional status as official observer. The fact that both *CB4* and *Fear of a Black Hat* share this structure and were in production simultaneously and independently makes the postulation regarding the rap-to-film transformation all the more viable.

Much the same thing happens in *Friday* (1995), a film starring and produced by rap's unrelenting Other, Ice Cube, who appeared in six films between 1991 and 1996, including *CB4*. And yet, unlike the persona used in his lyrics (and it is important to note that he does use a variety of personae), Cube's persona in *Friday* is not that of a gangsta other or a "natural born killer." Instead, he portrays a rather ordinary teenager living at home with his parents, dealing with parental pressure, the annoyance of his best friend's small-time scams, and ongoing harassment from this story's villain, a bully played by the hulking Tom Lister (formerly "Zeus" of professional wrestling fame). In other words, with this change in media came a change in character and ethos—he went from "the nigga ya love to hate" to the boy next door. The conflict with the bully comes to a resolution as Cube's character, heeding the guidance of his father, refuses to resort to gunplay (as does the protagonist in *Boyz in the Hood)*, and settles the score with the bully "like a man," in hand-to-hand combat. Witness also the reintegration of Ice-T to mainstream values: in his notorious *Cop Killer* (1992) album, he was accused of advocating sedition; five years later, he's been domesticated as a criminal turned lawman in the television series *Players* (1997). A loss of oppositional verve accompanied by narrative containment and commodification also affected the reception of the white Other of heavy metal. Initially viewed by the mainstream as destructive, in the Hollywood film heavy-metal musicians and fans are often portrayed as ineffectual, unintelligent, and benign comic figures, as in *This Is Spinal Tap* (1984), *Bill and Ted's Excellent Adventure* (1989), *Wayne's World* (1992), and *Airheads* (1994).

To provide one more example, another rapper, Tupac Shakur, whose gangsta persona has made him a rap hero, is assigned the role of demented villain in *Juice* (1992), while the story's hero chooses the good, middle-class road of hard work and achievement by making his mark as a rap DJ, Shakur's character, Bishop, is driven by a kind of gangsta idealism inspired by Hollywood (especially Jimmy Cagney in *White Heat* [1934]), which he formulates as a masculinist heroic ideal (and note that many of these films, as Diawara reminds us, revolve around a "ritual of manhood" [1993a, 24]). According to this heroic ideal, one can only transcend bourgeois social limitations through "bravado" and the necessity, as Dyson notes, of "remaking the world on [one's] own terms" (1992, 21). Dyson accurately notes a theme that reflects the notion of "retribalization" (see chapter 2), for the images of television have "replaced the Constitution and the Declaration of Independence as the unifying fictions of national citizenship and identity" (21). The hypermediation involved in this scenario is significant: Bishop, a gangsta character in the movie, is portrayed by the late Tupac Shakur, a rapper whose persona was deeply implicated in rap's "religion and ideology of authenticity" (Perkins 1996a, 20); but Bishop's source text is Hollywood gangster movies. They are the inspiration for his transformation into a "real" gangster, and thus authenticity itself is based on a mediation (according to this film, which is itself a mediation). Shakur's Jüngeresque military-masculinist idealism in evident in all of his productions, as in the introductory track of *Me Against the World* (1998), which, like the previously discussed Ice Cube album, begins with a fantasy of heroic self-possession in the face of death. This narrative and cinematic ethos is made all the more palpable by the events leading to Shakur's death in 1996. Vulnerable to charges that he was not a "real" gangster, Shakur played a role in escalating the East Coast/ West Coast rap feud, which led to his murder (Bruck 1997, 46–64).

These problematics are given further weight by *Menace II Society* (1993), a film that, while not dealing overtly with rap or rappers, uses a prominent gangsta rap soundtrack. The film's iconic center is the cruel and irrational murder of Korean merchants by a young black male, O-Dog, with the story's protagonist, Caine, along as an unwilling accomplice. After slaying the Korean shopkeepers, O-Dog takes the store's video surveillance tape to play for his friends. This reverses the situation in *Juice,* in which the character Bishop takes inspiration from old gangster movies and then goes on to perform a real act of violence. In *Menace,* the character O-Dog performs an act of violence, seizes the store's surveillance videotape, and thus becomes a kind of movie gangster for his circle of friends. This reversal shows how both films deal with the tangle of relations between mediated images, real acts of violence, and notions of authenticity.

But the very production of these films and their informing assumptions are also implicated in these mediations. The marketing of *Menace* as a work of social realism (the video box reminds us that "this is what's real") became a source of controversy and critique. bell hooks accurately hones in on the problem of reality and representation illustrated here. Drawing on James Lull's observation in *Media, Communication, Culture* that dominant ideologies work by transforming themselves into commonsense assumptions, she notes that

> even though in "real" life there is little evidence that young black males brutally slaughter Korean shopkeepers, when a film like *Menace II Society* depicts such a slaughter many young black folks insist that the dehumanized images of black masculinity are authentic, [that they] reflect reality. They both identify with and then seek to express culturally the identity the film gives them. (hooks 1996b, 75)

An interview with the film's directors, Albert and Alan Hughes (included as a trailer with the videotape), expands this problematic even further. When asked for an account of their cinematic influences, the directors responded that their two greatest influences were

Brian de Palma and Martin Scorsese (both noted for their portrayals of the Italian Mafia), and they specifically credit Scorsese's *Goodfellas* (1990), on which, according to the Hugheses, the beginning of *Menace* is based. And while the directors did bring in a number of "authentic" subjects for character modeling and technical advising and voiced some interest in showing how young criminals are created by their environment, the social realism of the film is largely forestructured by the codes of the successful Hollywood narrative. For instance, as revealed in the interview, they decided not to attempt a rendering of the actual social structure of the Los Angeles gangs (Bloods and Crips), which would apparently add needless complexity and inhibit the successful structuring of the film in terms of Hollywood's individualist ethos.

More revealing still is the construction of *Menace*'s protagonist. Boyd argues that, while John Singleton's *Boyz in the Hood* conveys a "bourgeois political position" because the protagonists, Tre and Brandi, get to pursue their middle-class dreams at Morehouse and Spellman Colleges in Atlanta, "the breeding ground for bourgeois Blackness, " *Menace* is characterized by a more nihilistic and marginalized viewpoint (1997, 97–98). But Boyd's interpretation is problematic, not only because of the Hugheses' self-professed admiration for mainstream Hollywood films but also because of the way the film deals with criminality and the criminalized Other—what the Hugheses refer to as "the underclass of a . . . people." The viewer's point of identification in *Menace* is through the protagonist, Caine, and while Caine, unlike the admirable Tre of *Boyz*, is at times a violent, impulsive, and irresponsible character, he is still a "middle" character, particularly when compared to his sociopathic friend, O-Dog. The Hughes brothers admit in an interview that they "had a lot of problems with Caine's character initially." They said, "[W]e went in and he was very violent character . . . we had to make him a more 'middle' character so that people could identify with him." While Caine, unlike Tre, remains unredeemed at the end of the story, what seems important in terms of the dictates of the codes of Hollywood is that he is *redeemable*, and while he is gunned down before he can leave Los Angeles with his girlfriend Ronnie (with her upwardly mobile aspirations), he was, at the time of his death, clearly on the path of redemption.

In its portrayal of black violence against Koreans, its use of videotaped violence as an integral element of the plot, its avoidance of a

portrayal of gangs, its use of authentic subjects for firsthand information, and its use of a "middle" protagonist-narrator, *Menace* demonstrates well the complex tangle of mediation, dominant aesthetic and
political ideologies, and reality. Interesting also is the transformation
of the Hugheses' professional reputation in the wake of *Menace*. Before *Menace*, when they were directing videos, they were only offered
rap videos; but after their directorial debut, they were offered video
work for rock and roll and other "white" genres, indicating that be-
coming a "film director" indicates some form of entry into the "white"
club.

The *House Party* films conform to a generic code that differs from
both the rap parodies and the films that purport to give a realistic
rendering of urban violence. They are closer in form to the domestic
comedy of the television series *The Fresh Prince of Bel Air* (1990–96). In
that series rapper Will Smith plays a West Philadelphia youth who is
taken in by upscale relatives in southern California and who "preaches
easy-going reconciliation with black elders and the black past" (Maxwell 1991, 3). The *House Party* series consists of four films (*House Party*
[1990], *Class Act* [1992], *House Party II* [1991], *House Party III* [1994])
starring rap duo Kid 'N Play (Christopher Reid, Christopher Martin).
When asked about his goals for the first *House Party*, Reginald Hudlin
expressed his annoyance with the fact that films about blacks have
generally been set in Harlem or Watts, thus ignoring "a wide range of
experience from people who live in Cleveland, Milwaukee, Kansas
City, Memphis; who live in houses or neighborhoods where you see
a cross section of classes and lifestyles." He said that it was this other
range of experience that he wished to capture in *House Party* (Hudlin
and Hudlin 1990, 66). It is not surprising, then, that the first *House
Party* films evidence a preoccupation with class affiliation in the black
community, for in each film the two rappers assume roles that effect a
quasi-essentialist reification of the class schism in the black community, with the lighter-skinned Kid taking the middle-class role while
Play portrays his less-advantaged counterpart, who tries to incline Kid
away from the middle-class values (education, hard work, responsible
behavior) of his father. This is also seen in the two lead female roles, a
light-skinned middle-class black and a dark-skinned counterpart whom
Play identifies as a "projects girl."

House Party 2 has a collegiate setting and can be viewed as a black acquisition of the college farce genre—a proven moneymaker since *Animal House* (1979). This particular generic affiliation is the source of the film's crossover power; however, it is also closely related to a more purely black American genre—the slave narrative, or what we might call the black *Bildungsroman*, in which a black protagonist (usually a young male) struggles against the irrationality of American institutions. The ur-text here is Frederick Douglass's autobiography (1845), with more immediate sources in Ellison's *Invisible Man* (1947), a text that has been referenced in other black films (cf. Diawara 1993a, 9) and that reverberates through *House Party 2* in a number of ways. Just as Ellison's narrator is sent by the good townsfolk off to college (a thinly veiled version of Booker T. Washington's Tuskeegee Institute), so too is Kid sent off to university, where he encounters for the first time the oddities of campus multiculturalism. For example, his roommate, Jamal, is a white boy who wears dreadlocks, keeps heroic portraits of Malcolm X and Nelson Mandela, and appropriates BVE (Black Vernacular English) so intensely that Kid pleads with him, "Jamal, do me a favor—talk white!" In an incident at the campus bookstore, Kid attempts a conversation with a fellow student, an Asian female. "And where are you from, Japan?" he asks. "No," she replies in standard American dialect, "Texas," and then delivers a sarcastic coup de grace: "[A]nd where are you from? Africa?" And thus he comes to learn, in *Bildungsroman* fashion, of his own preconceptions about others.

It is in the film's denouement we see more clearly the lineaments of class difference within the black community. Kid's counterpart, Play, dark-skinned and identified as more streetwise than book-smart, does not go on to university, but rather tries to convince Kid to eschew higher education in order to pursue success in the music industry. (The semiology of pigment here may evidence bell hooks's critique of the collusion with the hegemonic pigmentocracy that can be found in black American cultural production [1996a]). He is deceived by a black con artist (thus continuing the "black turncoat" motif first established by Douglass and other practitioners of the slave narrative) and thus gambles away Kid's scholarship check, which had been donated by the community's church. In order to keep Kid in school, Play makes a sacrifice: he sells his expensive car in order to replenish Kid's college fund. In this instance we witness a fictive attempt to heal the cultural wounds caused by the black class schism, for it suggests that the success of the black middle class, represented by the well-behaved and bookish

Kid, is predicated on the struggle and sacrifice of the underclass, represented by Play. The fact that his car, always loaded with eager young women, represents his sexuality (this is underscored by the car's name, as announced on its license plate—FOREPLAY) gives more meaning to his sacrifice, for it suggests the decidedly bourgeois precept that sexuality must be sublimated and channeled into disciplined intellectual labor. In the end, Play must settle for a battered and distinctly unglamourous car, which he names, significantly, FOREKID. And thus one class is sacrificed for the success of the other in an idealist scheme, for the good of all.

A more thorough exploration of this dynamic is found in the third Kid 'N Play film, *Class Act*, a comedy of errors in which bumbling high school administrators accidentally switch Kid and Play's academic records. Kid portrays here a young middle-class intellectual, an honor student who has transferred from another high school, with an absurdly highbrow name (Duncan Penderhughes), and his behavior is somewhat akin to the "black nerd" role popularized by the character Steve Urkel of the television series *Family Matters* (1989–)—apparently a new archetype, since it is not listed in Sterling Brown's comprehensive work *The Negro in American Fiction* [1937]). Play, meanwhile, portrays Blade Brown, a name that conjoins race with sharp criminality. Brown is an ex-con whose parole is contingent upon enrollment in high school, and thus, in this film class issues are rooted in the very premise.

The film engages the problems of black Americans as they try to cope with white-dominated institutions; after all, the failure of the white principal to detect the obvious differences between the two young men reinforces the Ellisonian theme of black invisibility. However, as Blade and Duncan attempt to work the role-reversal to their advantage, the plot soon abandons the problems imposed by American institutions at large in order to deal with the social problems within the black community in terms of class and socialization. Duncan learns to play the roll of hip-hop tough guy and Blade learns to act like a honor student from a solid upper-middle-class family; and it is no easy process, for there is considerable cultural difference between the two, perhaps best exemplified by the differences between standard and vernacular forms of spoken English.

"The theme of the comic," as Northrop Frye pointed out, "is the integration of society." In the form known as New Comedy, we are presented with an "erotic intrigue between a young man and a young woman which is blocked by some kind of opposition. . . . At the beginning of the play the forces thwarting the hero are in control of the play's society, but after a discovery in which the hero becomes wealthy or the heroine respectable, a new society crystallizes on the stage around the hero and his bride" (1957, 43–44). With the introduction of the two main female characters, *Class Act* begins to fit Frye's comic formula. Owing to the change-of-identity scam perpetrated by Blade and Duncan, they attract what at first seem to be inappropriate partners. On the one hand, a "projects girl" named Demeeta is attracted to Penderhughes (who she believes to be Brown); on the other hand, an upper-middle-class female named Ellen Grove (note the opposition of the two names, the one in keeping with a relatively recent trend toward invented, Africanized names, while the other connotes Anglophilic and bourgeois assimilation), whose parents own a business and attend formal dress cocktail parties, is attracted to Brown (whom she believes to be the intellectual and serious Penderhughes).

The parallel love affairs that develop give each of the male protagonists an incentive to truly learn, not merely imitate, the lifestyle of the other. The result is that Brown becomes a bit more sensitive, a bit less sexist, and he even comes to enjoy intellectual labor. Meanwhile, Penderhughes learns to appreciate the vitalism and energy of African American cultural forms (he was raised on European classical music). In one scene, Blade attempts to teach the awkward and clumsy Duncan how to dance and rap. As Duncan struggles, apparently trying too hard, Blade says, "[F]ifty percent is attitude and style. Body language is important, but the words"—Blade points to his head—"come from upstairs." If, as Frye suggests, comedy results in a "coming together," what the Greeks called a *gamos,* then such a process is clearly exemplified in this scene in which Brown teaches Penderhughes the codes of black cultural expression, which are composed of both "attitude and style" and, particularly in the case of rap, verbal skill and mental dexterity. By appealing to these intellectual qualities, which Duncan, as a good bourgeois, already appreciates, Blade has created an opportunity for a cultural *gamos,* with hip-hop serving as a locus for the reunion of two black American socioeconomic classes. And this cultural reunion is soon bolstered by biological union, for the love relationships that form between Blade and Ellen and between Duncan and Demeeta suggest a strengthening of black American culture by, again, a combination of working-class vitalism and middle-class values. And thus the film suggests, in narrative form, a reconciliation of the black proletariat (the mythohistoric guardian of the unique elements of black culture, hard won through bitter experience) and the black middle class (who have successfully learned to mitigate that bitterness through financial power and familial stability, but whose very success has, in either fact or appearance, ironically moved them from the cultural center of black life).

Class Act concludes with a *gamos*—a symbolic closure of the black American class schism— when, during a high school dance, Penderhughes takes the stage and shows that he too, like Blade Brown, has mastered the rap form, while that form itself, which now occupies the territory of an American middle-class institution (a high school) has come into the mainstream. Whether discourses like rap, which begin at the margins, can cross these lines of class, institution, and media with their authenticity intact, whether the concept of authenticity has any validity at all, and whether critics should concentrate on the instances of collusion and complicity in these cultural products or praise the emergence of black images that are at least to some extent black artistic control—these are questions beyond my scope here.

6

The Narrative Imperative

Four Exemplary Tales of Popular Modernity:
Collins, Kafka, Wrestling, Reagan

On 30 January 1925, Floyd Collins—a renowned spelunker who had discovered Crystal Cave in 1917—crawled into the narrow shaft of one of the thousands of limestone caves near the town of Sand Cave in central Kentucky. While crawling through a particularly tight passageway, Collins inadvertently loosened a large rock, which fell and pinned his left leg. He tried desperately to free himself, but only succeeded in causing more rubble to fall, covering all but his right arm, chest, and head. The next morning the accident was discovered and local authorities organized a rescue team. The story of Collins's predicament spread rapidly through the central Kentucky region and attracted the interest of the two competing newspapers in Louisville, the nearest metropolitan center. The *Louisville Courier-Journal* sent a young reporter, William "Skeets" Miller, who, upon being told in the early stages of the rescue attempt that if he wanted a story he could go down in the cave and get it himself, took the advice to heart. According to a fellow journalist, Miller, who was soon getting almost as much press as Collins himself, "took off his coat and rolled up his sleeves [and] got a talk . . . with the imprisoned man. Helped him. . . . Tried to get him out. Worked night and day. Risked his life. Risked pneumonia. Went on exhausted but determined, rescuing with one hand and reporting the story with the other" (qtd. in Murray and Brucker 1979, 133). By 3 February, the Collins case was the most prominent news feature in the United States, appearing in all the major newspapers and making the front page of the *New York Times* for five consecutive days.

Meanwhile, the rescuers worked on. They tried to dig Collins out, but it couldn't be done without risking another cave-in. They tried pulling him out with a harness, but quickly concluded that this would probably kill him. Then they considered amputating the trapped leg, but the shaft was too narrow to allow such an operation, and it was supposed that even if it were possible, he would probably bleed to death before he could be pulled through the narrow passageways and out to the surface. Finally, they decided to dig a parallel shaft in the hope they would reach Collins in time. One reporter summarized these efforts thus: "The pitting of the puny strength of man against the mighty forces of nature and the onward rush of precious time is one of the most thrilling spectacles ever given man to witness" (qtd. in Murray and Brucker 1979, 113).

And witness they did. Having either read about the incident or heard about it on Louisville radio, they descended on central Kentucky by the thousands—a responsible estimate puts the crowd on Sunday 8 February, which came to be known as "Carnival Sunday," at thirty thousand. Murray and Brucker recreate the scene thus:

> It was like a country fair, with a religious revival thrown in. . . .
> Families arrived in flivvers, clutching supplies of corn pone. Chil-
> dren spilled out and raced away as frantic mothers called for them to
> come back. . . . Lunch wagons appeared as if by magic, selling hot
> dogs, hamburgers, popcorn, pie, and apples. . . . Professional hawk-
> ers, meanwhile, offered all sorts of merchandise for sale—camp chairs,
> games, artificial flowers, cave onyx. . . . a favorite souvenir was a
> blue balloon on which was printed the words SAND CAVE. Hardly
> a child went home without one. (171–72)

In addition to this commercialism, the event had a "spiritual" element. As it was Sunday, clergymen took special advantage of the event,
not only in central Kentucky but throughout the nation. On that day
America's churchgoers were treated to sermons with titles like "Caught
in a Cave" and "Repent, Before It's Too Late" in which Collins and his
predicament became a convenient example of "sin cursed man . . . trapped
by the rock of evil" who can be freed and delivered unto the Lord
only through the power of prayer (qtd. in Murray and Brucker 1979,
172). The efforts of both estates (the church and the press) dovetailed

to create a national dramatic narrative as newspaper reporters offered the public their own belletristic versions of the special unity service held by the popular evangelist James Hamilton at the mouth of the cave, near the spot where geologists and miners were desperately trying to drill a parallel shaft through to Collins, who was now in more desperate trouble owing to a second cave-in. Hamilton's was, American newspaper readers were assured, "America's most dramatic sermon of the day. . . . The shrill screams of the steel drills in the pit below were his commas, the roaring concussion of the dynamite made his periods" (qtd. in Murray and Brucker 1979, 173).

As the days rolled by, new and sensational elements were added. Some reporters advanced the notion that Collins had been murdered; others claimed that food and water were purposefully being withheld from him so that he would die. There were also those who maintained that he was alive, but that he was able to come and go at will through a secret passage, and that he left the cave every night and returned in the day in order to continue a hoax whose sole aim was that of stimulating the tourist trade. But all speculation came to an end on Monday, 16 February, when the parallel shaft was completed and at last they reached Collins—or rather, his corpse. That same day, the *New York Times* announced the end of the rescue attempt with massive coverage, covering much of the first two pages and including six photos, including an old family picture of the Collinses that showed the young Floyd. In the wake of the failed rescue attempt, there were charges of incompetence that led to a well-publicized state investigation (Murray and Brucker 1979, 181).

Later that year, a blind Atlanta evangelist, Andrew B. Jenkins, teamed with singer Vernon Dalhart, famous for his folk ballads and "event songs," including "The Wreck of the Old '97," to record "The Death of Floyd Collins," which sold over three million copies in two years, far surpassing all other country record sales. Six other country singers also recorded cover versions (248). The ballad has Collins going off to the caves on that fateful day in spite of his mother's pleadings and his father's advice. It goes on to tell of the efforts of the rescuers—but of course they're too late. The ballad ends with a direct address to listeners, warning us to learn from Collins's example and to "get right" with God before it's too late (250–51).

There is a bizarre postscript to the story. Four years later, Collins's body was stolen from its tourist-oriented shrine: a glass-covered, bronzed coffin. The reporters once again descended upon central Kentucky, and a second rescue attempt was initiated—this time to recover an already dead Floyd Collins. The corpse, the left leg mysteriously missing, was found in a gunnysack near on the banks of the Green River. It was restored to its place; the thieves were never apprehended (235–36).

But it still wasn't over. As Carpenter's research demonstrates, not only did William Faulkner write an impressionistic poem based on the Collins case, which he probably learned of through the accounts in the *New Orleans Times-Picayune,* there is also some evidence that his novel, *As I Lay Dying* (1930), a comic journey to return a corpse to its proper resting place, may have been partially inspired by the Collins's story. Faulkner's novel, says Carpenter, constitutes a "macabre carnival . . . born of futility and the self-interests of those involved" (Carpenter 1995, 15). The Collins case cropped up again nearly three decades later in a screenplay and film by Billy Wilder, *Ace in the Hole* (1951). The story is set in New Mexico, with Kirk Douglas playing the role of a down-and-out reporter who, upon being asked to cover the story of a man trapped alive in a cave-in, recalls the Collins story, and then forms a partnership with a corrupt local sheriff facing reelection. Both journalist and politician hope to advance their careers by drawing out the rescue attempt longer than necessary in order to advance their own careers. Wilder, noted for his cynicism, was particularly interested in one aspect of the Collins case—the rumor that the entire incident was a stage-managed drama from the outset, and his screenplay is an imaginative development of that scenario (making this one of the first "antimedia" films, which, ironically, would become a stable commercial genre, as in *Network* [1976] and *Quiz Show* [1994]). Wilder's acerbic attack on media gone wild was a commercial flop. Madsen records that in an effort to reduce the massive resistance to the film, Paramount renamed it (*The Big Carnival* [1951]) and "took to sending relays of persuasive press agents around to city desks to explain that the picture's spectacle of trashy 'yellow' journalism was not directed against the Fourth Estate as such" (1969, 93). Journalists, including film reviewers, didn't care for Wilder's unflattering portrait of the news profession, and they returned Wilder's attacks by panning the film, thus proving Wilder's point about the power of the media just as the media corruption the film described became increasingly mundane—"standard operating procedure," as Gabler says in his remarks on *Ace in the*

Hole (1998, 82). (It must be said, however, that although the audiences of the day were not receptive to Wilder's cynicism, the film ultimately received critical recognition [Dick 1996, 61].) As for the Collins story, it has not entirely withered from public memory. In 1989 a Floyd Collins museum was opened as part of a bed-and-breakfast operation, and in that same year the Collins family finally gave him a proper burial—his body had remained in a chained, glass-topped casket near Crystal Cave since after it had been recovered after the body-snatching incident. And if you can't make the trip, you can see a photograph of Collins's tomb on *Roadside America* on the Internet (Kirby, Smith, and Wilkins 1996–97). Finally, in 1994, the American Musical Theater commissioned a musical by Adam Guettel and Tina Landau, entitled *Floyd Collins*, perhaps an appropriate venue for a revival of the first modern media event, which was theatrical in the first place.

There is a literary parallel to all this. In 1923, two years before the Sand Cave Tragedy, Franz Kafka wrote a "parable" entitled "A Hunger Artist." The story begins with Kafka's narrator bemoaning the problems faced by independent entertainers: "[I]t used to pay very well to stage such great performances under one's own management, but today that is quite impossible. We live in a different world now." And so it is that the hunger artist and his impresario become a subsidiary of a traveling circus—in many ways a prototype for modern corporate-managed entertainment. The impresario, realizing that it is no longer feasible to allow his client to simply practice the art of fasting, attempts to make him come to terms with some basic principles of advertising psychology and market demand by placing a forty-day limit on the fast, for "experience had proved that for about forty days the interest of the public could be stimulated by a steadily increasing pressure of advertisement, but after that the town began to lose interest." In addition to managing the public's interest, the impresario must also insure that the hunger artist's "performance" culminates in a spectacle, including a "flower-bedecked" cage, a military band, doctors to make an official announcement of the results through a megaphone, and two young women, who, "blissful at having been selected for the honor," help the hunger artist make his way to a small table laden with "a carefully chosen invalid repast." In spite of the impresario's efforts to make the hunger artist "interesting," however, in the long run his insistence on pursuing his craft beyond the limits of the public's attention span, combined with the competition provided by more kinetic and therefore exciting entertainments, brings the art of fasting to an end. The hunger artist is replaced by a young panther—full of vitality, a seeming joy for life, and above all, an appetite—who turns out to be a real crowd pleaser:

> [T]he joy of life streamed with such ardent passion from his throat that for the onlookers it was not easy to stand the shock of it. But they braced themselves, crowded round the cage, and did not ever want to move away. (Kafka 1983, 268–77)

At about the same time that the "Sand Cave Tragedy" was being produced and deposited into the general public mythology and Franz Kafka was writing his parables in the Prague ghetto, half a world away from North America's emerging media circus, something peculiar was happening in the world of sport. Although, like boxing, professional wrestling was, in the nineteenth century, a brutal sport well outside the margins of bourgeois respectability, it was, occasional theatrics aside (as under the management of P. T. Barnum), an authentic athletic competition. After the end of the First World War there were a series of legitimate wrestling champions, most notably George Hacken-schmidt, and in the 1920s a new generation of athletes and promoters came to the fore. The 1920s were characterized by a new intensity in terms of entertainment marketing, and it was an important decade in terms of the establishment of popular modernity's image-based con-sumerism. Professional wrestling adapted to these changes, and thus the new promoters and athlete/performers were primarily interested in reaping large profits. At this juncture, wrestling's inherent problems as a form of spectatorship had to be decisively addressed.

First, the problem of time: professional matches in the late nine-teenth century were often ten or more hours in duration, and three-hour matches were not unusual. Second, the problem of kinesics: in these authentic matches, as Morton and O'Brien point out, the audi-ence generally saw little more than two stationary figures locked to-gether at the center of the ring (a rather tiny ring for most of the audience in a large auditorium) moving through a series of relatively stationary "holds," the term itself clearly indicating the static nature of the contest. Furthermore, as Morton and O'Brien describe it, the moves and actions of a legitimate match are, unlike boxing, difficult to ren-der discursively. Radio provided a boost for many professional sports—

particularly baseball, whose spatial configuration is tailor-made for radio (easily held in the imagination and easy to describe spatially through fixed points and subfields—"third base," "left field," etc.) just as football is "made" for television (the shape of the playing field matches the screen space; plenty of "time-outs" to accommodate commercials). But radio was of little use for professional wrestling. The problem of wrestling is a problem of dramatic presentation; as Roland Barthes observes, "[A] boxing match is a story which is constructed before the eyes of the spectator; in wrestling, on the contrary, it is each moment which is intelligible, not the passage of time . . ." ([1957] 1977, 15–16).

For these reasons, a gradual transformation took place that destroyed the authenticity of the sport while allowing it to survive commercially. Wrestling became a staged spectacle rather than a sport by transforming its static elements into dramatic visual form. As Barthes explains, the gestures of professional wrestlers are "excessive . . . exploited to the limit of their meaning," and when the moment of victory and defeat is reached, these dramatic properties are maximized:

[A] man who is down is exaggeratedly so, and completely fills the eyes of the spectators with the intolerable spectacle of his powerlessness. . . . The gesture of the vanquished wrestler signifying to the world a defeat which, far from disguising, he emphasizes and holds like a pause in music, corresponds to the mask of antiquity meant to signify the tragic mode of the spectacle. (16)

The defining image, then, of twentieth-century professional wrestling is one of grotesquely exaggerated human suffering. And thus, during the 1920s the character of professional wrestling was indelibly altered. It became severed from both American amateur and the Greco-Roman traditions and was remolded according to the codes of an aestheticized entertainment spectacle.

Wrestling was not the only sport whose presentation as entertainment led to odd forms of transformation and mediation. In the early 1930s a peculiar practice developed, that of having radio announcers "re-create" a baseball game. The task of these announcers was to receive telegraphed play-by-play information from a live baseball game and to breathe life into it by reporting it with all the appropriate excitement, as though the announcer were actually there. We noted earlier that the telegraph was originally a handmaiden technology to journalism. Here we see that such a practice continued into the 1930s, but with the telegraph as handmaiden to radio journalism rather than to print journalism. It is curious that the counterfeit nature of these "re-creations" was widely known, and no one seemed concerned that a kind of fraud was being perpetrated. On one very interesting occasion, the telegraph line failed, and the person at the microphone was forced to improvise a baseball game that had no counterpart in reality whatsoever. The year was 1934, and that person was Ronald W. Reagan, before his acting career and long before his election to the presidency of the United States.

At the outset of his career in media and entertainment, Ronald Reagan found employment as a "re-creator," a task at which he excelled (occasionally an audience was invited into the studio to watch him work), probably because one of his first decisions as a media professional was to imitate the authoritative voice of radio announcers. Recalling the incident in which the telegraph went dead, Reagan says that it was very important that the radio audience was kept from knowing (Wills 1988, 119). The anecdote is instructive in terms of the relationship between illusion and reality in the Age of Reagan, and significantly enough, Reagan's approach to authenticity in this instance is related to at least one strand of public sports entertainment during his presidency, for the Reagan era is also the wrestling era.

In the 1950s (when Reagan was leading a union, the Screen Actors Guild, rather than going after one, as he did in the case of the air traffic controllers), the advent of television served to intensify wrestling's histrionics, particularly the tendency to emphasize personalities (such as the flamboyant Gorgeous George) and a hero/villain modality. Television introduced the now familiar wrestling staple, the between-bout interview, thus adding an dimension of discourse to what had hitherto been, in essence, a mime show pitting good against evil. After a period as a prime-time television event, wrestling fell into disfavor, and by the 1960s was strictly weekend fare. It appeared principally on the new UHF stations and served as an advertisement for the live event at regional arenas (Morton and O'Brien 1985, 48). In the late 1970s and early 1980s, however, wrestling experienced a major revival, helped largely by the emerging cable television industry and a format that was in fact a "slick . . . imitation of . . . weekly cable review shows in other sports such as football and basketball . . . [with] action highlights instead of whole matches from the various alliances around the country" (53). However, while it is true that the televised event became more "slick," a more important change was responsible for reinvigorating a moribund pseudosport. In the first shift in the 1920s, wrestling metamorphosed from competitive sport to theatrical ritual. In its second shift in the 1980s, the theatrical elements were expanded: just as the stage-managed events often spill beyond the contents of the "squared circle" so that the participants carry out their scripted roles on the concrete floor of the arena, so too do the narratives in modern wrestling go well beyond the matches themselves, thus turning a series of matches into a larger story, with each match serving as a battle in a larger war. As Leland notes, these story lines now overshadow the wrestling itself (1999, 62).

This element was already present to some degree in televised wrestling in the 1960s, particularly in the interview segment of the program in which wrestlers issued challenges to their foes (known as "cutting a promo"). Another common pattern was for the villain to "pearl harbor" and injure the hero during the interview, thus postponing the final delivery of justice (and justice is, to a large extent, wrestling's central theme) until the live event. Yet another plot involved the pitting of a hero dedicated to "scientific" (i.e., fair and rational) techniques against a villain ever willing to use underhanded methods and the assistance of his henchmen as the referee stands helplessly by (Morton and O'Brien 1985, 106). In this way, the limited morality play of the single wrestling match takes a serial form and emphasizes what sociologist Thomas Hendricks rightly identifies as one of the major themes of wrestling: the "display of official incompetence" (1974, 184). In the morality play of wrestling, the referee's role is that of fool. In an ironic inversion of Blind Justice, here blindness is not a sign of impartiality, but of ignorance.

The "Wrestlemania" events (established in 1985), as Twitchell reveals, were elaborately scripted and densely intertextual, bringing entertainers, television characters, and "real" political figures into the same spectacular narrative construct. In the first "Wrestlemania," Hogan teamed with Mr. T of the television series *The A Team* (1983–87) to settle a grudge match with villains. This event had at its center a bloodfeud narrative that had been carefully managed over the course of months through the Worldwide Wrestling Federation's staggeringly successful cable television contracts. On hand for the event were Muhammad Ali, Liberace, Billy Martin, "New Wave" rock star Cyndi Lauper, and Gloria Steinem (who added a feminist element by calling the kilt-wearing Scotsman Roddy Piper a "disgrace to the skirt"). And as a final coup de grace of pop culture expansionism, vice presidential candidate Geraldine Ferraro also entered the game by calling on Piper to "shut up and fight like a man" (qtd. in Twitchell 1989, 249).

By the mid-1980s, professional wrestling was America's fastest growing sport ("Wrestlemania II," at the Silverdome in Pontiac, Michigan, set a new record for the most well-attended indoor sporting event in modern history). The Wrestlemanias that followed stuck fairly close to the plots and themes established in the 1985 event. "Wrestlemania X" in 1995, for instance, featured a match between Owen and Bret Hart billed as a "brother against brother" blood feud; the promotional buildup to the event described how the Hart brothers' mother was so distressed by this feud that she had to be hospitalized. Ironically and tragically, in 1997 Bret Hart would publicly denounce the WWF for going too far with the antiheroes like Stone Cold Steve Austin (cf. *Hitman Hart* 1998; Leland 1998), and in 1999 his brother, Owen, would die in a wrestling stunt gone wrong. Other recent (1990s) innovations include the "porn star" wrestler, Val Venus, and the Nation of Domination, a league of evil wrestlers whose entourage includes gangsta-style rappers and suited-up bodyguards in parody of the Nation of Islam. This continues wrestling's tradition of evil ethnic Others, often used to appeal to the crowd's nationalism. Its particular emphasis in the 1970s and 1980s on evil Middle Easterners (e.g., The Iron Sheik) once again shows the connection between geography pedagogy and racial typecasting.

In its contemporary form, wrestling has a narrative dimension that, as Jenkins observes, goes well beyond what Barthes could have predicted, taking the shape of a masculinist soap opera (1997, 49–50). In many ways, the characteristics of professional wrestling make it a prime example of popular modern entertainment.

Popular Modernity and Narrativity:
Experience, Technology, Mythohistory

The chapters that make up this book have been organized in part with reference to experiential categories—the experience of spatiality, temporality, the image, and the human voice. In this, the last chapter, we have turned to narrative, but it is not at all clear that narrative can be regarded as an experiential category. Narrative, first of all, is not exclusively associated with any particular sensory mode, as are listening and looking. We see spatial fields and images with our eyes, we hear voices with our ears, but we can experience narrative in a number of ways: a narrative can be told to us by a speaker, or it can unfold through pictures; and in written narratives, we "hear" the voices of speakers and we "see" mental imagery. Some literary forms (Pound's imagism, or what M. M. Bakhtin identifies as the polyvocal quality of a Dickens and Dostoevsky) use these features as a specific aesthetic strategy. In this regard, narrative is closest to our experience of time, not only because it operates in a chronological framework but because while we speak of "watching" or "hearing" the passage of time, our temporal awareness is not necessarily tied to these sensations (although it probably has some relationship to the "biological clock" of cellular processes).

An interrogation of narrative vis-à-vis experience can be found in several locations within the hermeneutic phenomenology of Paul Ricoeur, beginning with *The Symbolism of Evil*. In his analysis of a particular narrative practice, the confession, Ricoeur finds it first necessary to interrogate the "zones of emergence" of symbolism; the manner, that is, in which a language practice springs forth from experience. He reasons that, in some originary historical experience, the human being first finds the sacred in the physical world or in some specific aspects thereof (some "fragment of the cosmos"), which, assumably through an intentional act, loses its "concrete limits" and becomes charged with meanings. Drawing on Eliade, Ricoeur argues that the emergence of the sky as symbol demonstrates that in this originary example there is no disjunction between world and sign. The sky manifests those things that it will ultimately signify—from the "most high," the "elevated," and the "immense" to "the clairvoyant and the wise, the sovereign, the immutable." In this catalog we can indeed trace the semiogenesis of the transcendental signified from the experienced world.

It is at this juncture that we must return to our considerations of the human voice: this experience of the symbolic, of an invisible cosmic principle manifested in the perceivable world, Ricoeur agues, "gives rise to speech" (1967, 10–11). This emergence of speech, he insists, precedes that of thought, and we may conjecture that it is a response to the (hypothetical, primal) subject's confrontation with the transference of experience into symbolic meanings. Perhaps also, the emergence of vocalization in the face of the symbolic experience is a mimetic act, for elsewhere, and again with reference to Eliade, Ricoeur says that the sky "*speaks* of wisdom and justice by virtue of the analogical power of its primary signification. Such is the fullness of the symbol as opposed to the essential emptiness of the sign" (1974, 319; cf. Klemm 1983, 67–68).

Vocalization as a response to the symbolic reflects an attempt to imitate the quality of the symbol as it appears in nature, a response to the powerful stimuli of the moment of symbolic apperception, and an attempt to speak, like God, from the sky (or the mountain, or the whirlwind). Finally, there is a temporal element here: the voice that speaks speaks over time, and perforce speaking becomes narrating. And thus, not surprisingly, in the later development of his theory Ricoeur posits a primarily temporal hermeneutic function for narrative: both historical and fictional narrative, he argues, negotiate the *aporia* that separates phenomenological time (the time of lived experience) from the unknowable cosmic time of the world (1984–85, 99, 104).

Based on the above sketch, then, we can offer a postulation regarding narrativity and experience: narrative acts as a coordinating structure—it coordinates primary sensory experience (sound, sight, and the primary experience of spatiality) as it emerges into symbolic experience, the voice that comes out of this experience (the transference of experience into expression), and, finally, the less sensory-oriented experience of the temporal. What is being coordinated, then, is voice, spatiality, image, and time: the experiential categories that have guided our examinations throughout this book. Reflecting back, then, on where we have been, let us scrutinize this postulation regarding narrative.

We begin with the voice, for here the connection with narrativity is perhaps most obvious; in fact, according to the above account, voice is a kind of threshold between symbol and narrative. In the second chapter, we focused on the voices of broadcast radio in the 1920s and 1930s—specifically, the voices of Charles Coughlin, Franklin Roosevelt, and Bing Crosby—and then, in a later discussion in chapter 5, the vocal ethos of rappers like Ice Cube. In these analyses we observed the particular association of the voice with authority, and it was suggested that the function of the superego is that of providing a mental-aural archive of the disembodied voice of the father. Here the disembodied voice of popular modernity has powerful atavistic associations, an observation that picks up some cultural-historical validity when we recall the speaking tubes of the ancient priests referred to by Jaynes, or, for a closer example, the disembodied voice of the priest (father) on confessional duty. But these groundings go even further when we add to them Ricoeur's notion of the world "speaking to" man. If this be so, then the function of the narrator's voice is a metonym for the voice of the world. The narrator of Genesis is the human voice that takes the place of the voice of the world, adding to that original voice sequence, causality, explanation—in short, narration (interestingly, in Genesis, God speaks, but it is the human being, Moses, who narrates).

FDR's effectiveness with radio was, as we noted, rooted in his intonation and his rhetoric (particularly "radio rhetoric," as Cantril and Allport call it, which produces an impression of intimacy), but we did not adequately indicate the role of narrative. Political discourse in constitutionally based, allegedly rationalist societies is purportedly more invested in exposition than narration. FDR's first Fireside Chat, on the surface of it, is correspondingly more expository than narratorial. FDR explains how banking works ("let me state the simple fact that when you deposit money in a bank the bank does not put the money into a safe deposit vault. It invests your money in many different forms"); he explains why there is a "bad banking situation" and how that situation came about (bankers who were "either incompetent or dishonest in their handling of the peoples' funds"); and he explains what he intends to do to solve the crisis (61). The speech's narrative structure is more rudimentary: it is a chronologically ordered outline of the steps that led to the proclamation of a nationwide bank holiday and the legislation that followed, which gave him the authority to effectively deal with the situation. But the reference to a "bad banking situation" and bankers who were "either incompetent or dishonest in their handling of the peoples' funds" reveals a deeper narrative structure (Barthes's "second-order" system of myth, which we'll get to shortly), one that is more fully revealed in a speech that he delivered only a few days earlier: his "Inaugural Address" of 4 March 1933. In this speech, no doubt still fresh in the minds of those listening to the second ("banking crisis") speech, FDR, in the first few sentences, sounds a rationalist note by suggesting to the American people that they turn away from "fear itself—unreasoning unjustified terror which paralyzes needed efforts to convert retreat into advance" ([1933] 1938, 11). But paradoxically, FDR soon turns to counterrationalism by converting that fear into anger. As Ryan notes, the ancient scapegoating technique is leveled squarely at America's financiers (1988, 79). FDR charges these "rulers of the exchange of mankind's goods" with failure, incompetence, dishonesty, greed, and lack of vision. These "money changers," FDR states, have "fled from their high seats in the temple of our civilization. We may now restore that temple to the ancient truths. The measure of the restoration lies in the extent to which we apply social values more noble than mere monetary profit" (1938b, 12). In short, he charges that the financier class is uncivilized; they are ignorant of the social goals of civilization and have resorted to a primal, presocial, and even sacrilegious greed. As Colunga puts it, in this speech FDR posits a "combatable enemy" and "lays the blame for the poor economy on the doorsteps of the banking and finance industries," and as he does so, he sets himself in the role of a savior. He "speaks in a decidedly religious vein," with the use of biblical reference (money changers, temple) enhancing the religiosity of the speaker and promoting the sense of his godliness (1993, 51–52). Here then, we see the narrative underpinnings—the same underpinnings we saw in the civilization:savagery binary of Frederick Jackson Turner's frontier thesis (or perhaps we should say, frontier mythos). The purportedly rationalist exposition of the modern politician-statesman, as with the empiricist historian, is the prerational narrative of good versus evil.

In our discussion of spatiality, we noted that maps form an instrumental technology whose use-value is that they present representations of spaces beyond our immediate apprehension (the *Ursprung* here being cosmic space and the experience thereof)—but in so doing, they also narrate. In the mapmaking practices of the seventeenth-century American colonists, native people were generally represented as marginal decorations. In a sense, then, these maps told a story about Europeans and Native Americans: the story, or rather myth, of Manifest Destiny. But this is not only true in the case of these extraordinary antiques; as Denis Wood demonstrates, even a map as prosaic, contemporary, and unassuming as the "North Carolina Official Highway Map, 1978–1979" has narratological features. Using Barthes's notion of the "second-order" nature of myth's semiotic functioning (i.e., a story not directly told but understood on some level) ([1957] 1977, 114–15), Wood reveals how the text and icons of the map's legend mobilize to tell a public relations-oriented story about North Carolina, a state, the hypothetical map reader is to conclude, that is all at once a "*hip* state (though bucolic), urban, urbane, sophisticated (but built on traditional values)" (1992, 102). In short, maps narrate.

In chapter 4, we dealt with popular modernity's uses of the image and the icon. In the gloss of Ricoeur, we encountered the notion that the symbol is based on some perceivable image that becomes "charged with meanings," and though this perceivable external thing is now rendered as a content of consciousness, first as mere sensation and then as a meaningful symbol, it nevertheless retains a kind of primary association with its external, if you will, "mere-thingness." The power of the image-symbol, then, like the power of the voice, is rooted in this primary activity. The power of the image-symbol is such that, with the voice, it remains central to both literary and popular narrative; perhaps the need for verbal imagery in text-based narrative is a way of counteracting what Ricoeur calls the "emptiness" of the sign with the "fullness" of the symbol.

In any case, one element of the image-narrative nexus is this quality of image embeddedness in the aural-textual frame of the narrative text. For an example of this in popular modernity's narrative production, we need only return to the Sand Cave Tragedy, which has at its center an image of considerable power—that of a man trapped and shuddering in the cold dark earth. It was the horror of the image of Collins's situation, not the newsworthiness of the event, that caused the story to grip the national imagination for nearly three weeks. In one journalistic account, readers were told that "fiction has nothing comparable to the horror of this scene which makes one quail and shudder," and neither Poe nor Conan Doyle can tell stores "comparable to the throes of agony tearing the spirit of Floyd Collins" (qtd. in Murray and Brucker 1979, 155). This conjoining, in popular modern narrative, of entertainment with allegedly newsworthy "information" presented under the aegis of rationalism, will occupy us again shortly.

If it is true that the verbal texture of narrative is dependent on elements of imagery, the inverse is also true: that images can and do narrate without the overt presence of a narrative framework, with the narrative merely implied and imposed on the image-icon by way of the subject's cultural literacy. Here we note the two-way trafficking involved in the symbolic process: images can have a prenarratorial reference in the primary anthropocosmic activity described by Ricoeur, but they can also have a post-narrative reference. We already remarked on how this process can work in relation to ideological concepts in chapter 4 when we spoke of an image:concept ratio that brings an external ideology into the subject in the form of an image or a symbol.

It may well be that both processes (one based in Ricoeur's hermeneutic phenomenology, the other in a Marxian phenomenological psychology) need to be taken into consideration. Imagine, for instance, a Hallmark card of the "bereavement" genre, a card that features a photograph of the sky (sunlight streaming through clouds, some appropriate words of sympathy emblazoned across): is the experience of sky qua symbol dependent on the atavism Ricoeur describes, or on culturally received sentimentality? What does "sky" mean here, and what is the source of that meaning? If we think back to our discussion of male icons like Rambo, RoboCop, and the stars of bodybuilding and wrestling, we find similar complexities. The sheer bodily mass of these figures reminds us of Ricoeur's concatenation of symbol and thing, of the thing manifesting the symbolic property (in this case, physical strength), a perception that seems more the direct result of the image itself rather than any contemplation about it (as Ricoeur suggests, the symbol precedes the thought). On the other hand, the icon is imbued with cultural meanings. In the case of RoboCop, this cultural polyvalency is evident, first, in the notion of exaggerated masculinity (a cultural construct if ever there was one), second, in the myth of the man–machine, a standard part of modern cultural literacy since Mary Shelly's *Frankenstein;* and third, in the fascist metal-body aesthetic (a metonym for the social order). (And, as we noted before, the actual narrative of *RoboCop,* which has an antifascist and anti-corporate element, is not necessarily the informing narrative in the cognition of the RoboCop icon.) Another good example would be the semiotics of music. The sound of a musical instrument (like the violin, for example), as we noted before, presents itself as acultural; and yet, like the image of the sky, it comes to us with its own cultural baggage.

This two-way trafficking in beholding the image-icon, then, takes place along the chain of signification that arises from the symbol. We saw it in the sky being, first, the "most high" and the "elevated," and then later "the wise, the sovereign, the immutable." In the case of the male body icon, the movement is from "physical strength" to, say, "virtue, self-sacrifice, purity." But the point here is simply this: the visible symbol does not only manifest. It can also convey myth and, as a result, narrative.

* * *

Demonstrably, then, narrative coordinates sensory experience into a meaning structure, and having disposed of the question of the experiential status of narrative, at least for the moment, we can move on to the second element of the experience/technology/mythohistory triad. But rather than looking at technology as something other than narrative, it seems that much of what has transpired thus far indicates that narrative is not an "other than" here—rather, narrative *is* a technology. In his eidetic reduction of technology, Heidegger moves beyond instrumentalist definitions to come to the conclusion that technology is a "revealing":

> [W]hoever builds a house or a ship or forges a sacrificial chalice reveals what is to be brought forth. . . . thus what is decisive in *technē* does not lie at all in making and manipulation. . . . but rather in . . . revealing. It is as revealing, and not as manufacturing, that *technē* is a bringing-forth. ([1962] 1977, 13)

In common parlance, "technology" is taken to refer almost exclusively to modern technological instruments, anything from, say, a lightbulb to a mainframe computer (although the lightbulb, an example of what Marvin calls an "old technology," is so taken for granted that it barely makes the list). Windmills, irrigation canals, and slate roofs, to say nothing of adobe construction or Native American farming, hardly strike us as being technological at all, given our Eurocentrism and modernity-centrism. Heidegger's reduction first of all unburdens us from these constrictions, highlighting the creative aspects of technology rather than merely its instruments. Technology is seen as a human way of calling forth the power of nature to reveal itself, or a calling on nature to reveal the plans of a human maker, as the way various craftsmen and tradesmen call forth the architect's vision from nature, from wood and stone.

If we accept this technology-as-revealing and add to it the proposed narrative-as-technology, we are inevitably left with narrative-as-revealing. And so, what is it that narrative reveals? Let us first return to our provisional definition: narrative provides a coordinating structure for primary sensory experience as it emerges into symbolic experience; it is the voice that comes out of this experience (the transference of experience into expression), and, finally, the less sensory-oriented experience of the temporal. To argue through analogy, then: the carpenter and mason reveal something not present in nature, namely, the house, and beyond that, the home that was latent in nature and in the mind of the maker; the house is something more than the aggregation of its wood and stone. So too is a narrative more that its sensory elements. It is more, even, than the coordination of these elements into a structure, just as "home-ness" (the underlying human need that gives rise to construction technology) is more than "house-ness." The thing that is revealed by the narrative act is meaning. Or, as Polkinghorne puts it, narrative is a

> scheme by means of which humans beings give meaning to their experience of temporality and personal actions. Narrative meaning functions to give form to the understanding of a purpose to life. . . . It provides a framework for understanding the past . . . and for planning future actions. It is the primary scheme by means of which human existence is rendered meaningful. (1988, 11)

In this function of revealing meanings, we can say that narrative also reveals our location, what we might term a "meaning-locus," identified by the common phrase "my place in the world," which or course does not refer to physical emplacement at all. The narrative technology, as a revealer of location, is part of a family of technologies: the clock and the calendar reveal to us our location in time; maps reveal our location in space; narratives reveal our location in meaning, or, more accurately, what we believe to be our location in meaning.

And just as the question of narrative experience collapses into the "question concerning technology," technology in turn collapses into myth, for the meanings that narratives deliver are not empirical truths, nor are they whatever might be signified by the immediate narrative syntagam. They are, rather, not directly stated or even directly implied meanings of what Barthes calls the "secondary semiotic system" of myth ([1957] 1977, 114–15). Consider again the story of Floyd Collins. The events at Sand Cave, Kentucky, were certainly not of any real national significance, though they received more media attention than other cave-ins—mining disasters—involving many more lives, this in keeping with Lenin's dictum, "one death is a tragedy, a million is a statistic." Given its lack of national significance, not to mention its lack of pragmatic use-value to individuals in the course of their lives, we are left with the question of its prominence as a news story. We have already noted, in a discussion of the origins of the image-symbol, that the Sand Cave Tragedy centers on image—an image of horrific suffering. In accord with Ricoeur's notion of the primary formation of the symbol, perhaps the image at the center of the Collins story can be explained by the manifest quality of earth itself: Collins is a symbol of suffering, and ultimately, of death, because the image of him trapped in the cave is an image of death, of being quite literally swallowed up by the earth. But here what I have called the "two-way trafficking" of meaning-formation as a forestructure of received ideas comes into play. The very image of suffering is of course the culminating event of the Christian gospels, and it later became the absolute center of Catholic iconography (the popular culture of the baroque, as José Marvall argues); the Collins story resonates with elements of Christian mythic forms, and thus, any number of fundamentalist ministers relentlessly perused the Collins story as a metaphor (or in the Puritan tradition, a typology) for man's quest for salvation. Perhaps the Carnival Sunday spectators were hopeful that they would see Collins emerge from behind the rock, saved by, on the one hand, the force of God's grace as preachers invoked the metaphor of "sin cursed man trapped by the rock of evil," and, on the other, by man's own technology. The image of popular evangelist James Hamilton preaching at the site of the rescue attempt, his voice mingling with the sounds of the earth-moving machinery, leads us back to the notion that popular modernity is driven by what Slater calls "natural magic"—a blending of rationalist technology with prerationalist belief systems. But as Collins did not emerge as a resurrected hero, the story was reshaped, as in Dalhart's popular ballad, in accordance to the fundamentalist ethos connected with what Sacvan Bercovitch calls the tradition of the American jeremiad, exemplified by Jonathan Edwards's "Sinners in the Hands of an Angry God." The Collins story became a warning to (as the popular ballad puts it) "get right with your Maker," and its power as a public narrative was connected with the ideological struggle that occurred in the 1920s as a result of the reencoding of the United States as a more licentious consumer-oriented culture. The point is that the interest generated by the Collins story is a deeply structured, mythic one, and the mythic constellation here is the same one we find in the other threads from our skein of narratives: suffering, as Barthes points out, is at the center of the professional wrestling spectacle, and likewise, it is the central at-traction for both the circus spectators in Kafka's parable and the read-ers of the parable.

The American presidency as it relates to mythic narrative has a particular significance, for in the case of a national leader (whether president, dictator, or king), unlike that of a professional wrestler or a man stuck in a cave, the individual in question is merely the mortal, temporal embodiment of a larger entity. The symbology and mythos of the national leader is thus an example par excellence of Ricoeur's notion of the phenomenological and hermeneutic function of narra-tive in providing a third form of time, a mythic time, that mediates the larger, unknowable time of the cosmos and immediate, experienced time. In the case of the national leader, this is hardly a subliminal affair, as public discourse makes constant reference and clear distinctions be-tween "the President" and, usually in somber tones, "the Office of the Presidency" (this notwithstanding the Enlightenment project of de-mythologizing public offices). The cultural function of the presidency partakes of the mythohistorical because of the use of narrative to place an individual in a broader chronological frame (just as the pedagogical geopolitical maps place one in a broader spatial frame).

Arguably, this was all the more true in the Age of Reagan, an era when, as Combs notes, the functions of the presidential office were "less political than they were mythological" (1993, 33). We should not underestimate the mythical value of an incident that happened early in his presidential tenure—the failed attempt on his life by John Hinckley. Hinckley's failure enabled Reagan's heroic myth: as he lay wounded in the hospital, Reagan is reported to have quipped to his wife, "Honey, I forgot to duck," and thus the Age of Reagan began mythopoetically, with the hero who rises from suffering and to laugh in the face of death. But the Reagan image was a polyvalent one, combining this image of tough-guy hero with a nostalgic mythos. As Combs argues, Reagan came to national prominence as a candidate for the presidency at a time of accumulating national angst: the expe-rience of failure, both moral and military (Watergate and Vietnam), the loss of economic omnipotence through the rise of Japan, the Arab oil embargo, problems with unemployment and inflation, and, in gen-eral, a growing sense of national decline. Reagan, the most popular president in history, provided an antidote to these feelings by provid-ing a prelapsarian myth, one rooted in a Rockwellesque images of the small town, an image Reagan exploited skillfully in his 1984 presidential campaign during a well-publicized visit to a house in Dixon, Illinois, where he and his family lived as renters from 1920 to 1923 (28–35). Significantly enough, the popular films of the Reagan era, particularly the *Back to the Future* trilogy, are, as Nadel argues, heavily invested in the question of historicity, nostalgia, and the national past, themes that play again somewhat later in *Apollo 13*.

If, then, narrative is in essence a technology that coordinates experience and provides the subject with a meaning-location through the second-order signification that is myth, we might wonder as to the differences, if any, between narrative under popular modernity's intersubjective perceptual regime and that of any other period. To some extent, the mythological argument counters the historical one, but there is no reason to mount an exclusionary argument here; that is, the presence of transhistorical constructs does not invalidate or trivialize that which is peculiar to a given epoch, and in terms of popular modernity, one great particularity comes in the form of the technoeconomic and managerial structures of the secondary enabling technologies that create, distribute, and to some extent modify mythic narrative.

The entire system of production and distribution for mythic narratives qua entertainment constitutes a modern capitalist technology, and thus we might expect to find some parallel in the differences between, on the one hand, the practice of an oral storyteller and that of televisual narrative, or, on the other, between the technology of the windmill and that of the power plant. In his discussion of such differences—the differences between premodern and modern technologies—Heidegger notes that they are both "revealings," but with a significant difference:

> [T]he revealing that holds sway throughout modern technology does not unfold into a bringing-forth in the sense of *poiēsis*. The revealing that rules in modern technology is a challenging [*Herausfordern*], which puts to nature the unreasonable demand that it supply energy that can be extracted and stored as such. But does this not hold true for the old windmill as well? No. Its sails do indeed turn in the wind; they are left entirely to the wind's blowing [and] the windmill does not unlock energy from the air currents in order to store it.
>
> In contrast, a tract of land is challenged into the putting out of coal and ore. The earth now reveals itself as a coal mining district, the soil as a mineral deposit. The field that the peasant formerly

> cultivated and set in order [*bestellte*] appears differently than it did
> when to set in order still meant to take care of and maintain. ([1962]
> 1977, 14–15)

Heidegger's own nostalgia is something we could critique here, something that reminds us yet again that the critic can never fully detach
him- or herself from mythic narrative. But having said that, Heidegger's
distinction between these two technologies has considerable warrant.
We find some of these observations reflected in Lewis Mumford's
notion of the "megamachine." While Mumford is specifically interested here in the evolution of warfare, his remarks also relate to the
difference between old and new technology. The traditional military
state apparatus, Mumford says, was reliant on manpower, which is
subject to "attack from without and corruption from within." The
megamachine, on the other hand,

> knows no such limitations: it can command obedience and exert
> control through a vast battery of efficient machines, with fewer hu-
> man intermediaries than ever before. To a degree hitherto impos-
> sible, the megamachine wears the magic cloak of invisibility: even
> its human servitors are emotionally protected by the remoteness from
> the human target they incinerate or obliterate. ([1964] 1970, 267)

Are these observations relevant to mythic narrative in the era of
popular modernity? Certainly, Heideggerian and Mumfordian notes
clearly sound in those critiques and theories that are more specifically
aimed at postindustrial (or Fordist, to return to Harvey's term) rather
than industrial technology. For instance, Ross Snyder claims that in
"masscom" society, subjects are "not primarily rooted" in the natural
world, indeed, they "hardly experience this world of nature, let alone
[are] formed by it" (1979, 351). Similarly, Walter Ong claims tenta-
tively, "[I]t would appear that the technological inventions of writings, print, and electronic verbalization, in their historical effects, are
connected with and have helped bring about a certain kind of alien-
ation within the human lifeworld" (1977, 17). And then there is Jean
Baudrillard's notion of the "simulacra," a world in which one only
comes in contact with imitations.

We return again to our skein of stories to see how these criticisms
apply. The selection is disparate enough: the media's coverage of a
cave-in, a short story from the high modernist canon, a popular enter-

tainment spectacle, and an American president. In spite of this dispar-
ateness, or perhaps because of it, these examples reflect an essential
ingredient of popular modernity, namely, the movement towards a
engineered mass culture. Clearly, Kafka's "A Hunger Artist" stands as
an allegory for the establishment of the growing relationship between
the business disciplines (advertising, accounting) and cultural production as well as the general economic format of popular modernity,
which is paradoxically driven by consumer demand while at the same
time being manipulative of it. The impresario's embellishments are
wisely chosen indeed, for they combine elements that have nationalistic and militaristic associations (the military band) with those that convey sex appeal (the young ladies who lead the artist from the cage and
who, like the models on television's *The Price is Right* [1956–] and
Wheel of Fortune [1975–], have no other function than that of being
seen), as well as those that convey the imprimatur of official expertise
(the certifying physicians). All of the elements of this semiotic clutter
have appeared in our survey of popular modernity's iconography, from
the nationalism of *Independence Day* to the use of sexuality in all manner of advertisements, to the role of the physician's testimonial in the
kinds of ads that emerged in the 1920s for products ranging from Lysol
to Cream of Wheat, which employed a strategy that Marchand calls
the "democracy of afflictions." Further, the final failure of the artist
and his replacement by the young panther should lead us to recall
vitalist movements of the early twentieth century, from the various
programs in United States that formed what Lears calls the "therapeutic
ethos" to Marinetti's futurism, Italian fascism, and German Nazism,
and finally back again to the advertisements of contemporary consumerism, perhaps best exemplified by the advertisements for Pepsi
discussed previously. What all of these things have in common is the
valorization of youth, vitality, and athleticism, and their dispensation
as part of an engineered mass culture via the organs of the state, or
private industry, or of some combination of the two.

In keeping with Heidegger's theory concerning power plants and
mining operations, the impresario's managerial technology of using a
forty-day limit (a secondary *technē* that rests on the artist's primary "art
of fasting") is an attempt to exploit the entertainment spectacle to the
limit of its value, extracting every last morsel of human interest in such
events. The interest in the artist is no longer like the relationship of
wind and windmill—it is more that of a *Herausfordern,* a "challenging,"

an attempt to engineer the extraction of a maximum profit in terms of
ticket sales, just as the mining operation seeks a maximum yield.

We have spoken of the uncanny parallel of Kafka's parable with
"Carnival Sunday," the day that thousands of people massed at Sand
Cave, Kentucky, lured on by the strangely invisible "spectacle" of a
lone, suffering human being. From the lunch wagons and souvenir
hawkers that materialized "as if by magic" to the preachers who shaped
their sermons around the events at Sand Cave, to the journalists who
competed to get the information first and to uncover "the real story,"
the entire episode in Kentucky reveals how, under the regime of popular
modernity, random events quickly trigger the swinging into action of
a waiting information management apparatus (another a good example
would be the 1996 O. J. Simpson murder trial). Had he lived, perhaps
Collins, like the hunger artist, would have needed an impresario. That
is precisely what was suggested by one of the founders of "image management," E. E. Calkins, who commented on the Collins phenomenon three years later:

> If poor Floyd Collins, who died in a hole in Kentucky, had been so
> fortunate as to come out alive, he too would have needed a press
> agent and an attorney, even more than a doctor, because he would
> have found that Kentucky cave a gold mine of nation-wide public-
> ity, which could and would be exchanged for large checks in pay-
> ment for services for which he had no other fitness than that his
> name was known to millions. (1928, 193)

Certainly, one could scarcely invent a more apt illustration of popular narrative practice vis-à-vis technology than the Sand Cave Tragedy, which was one of the major media events of the 1920s and a
foreshadowing of the power of the emerging web of communication
and information systems. The speed at which the event was broadcast
throughout the United States was one of the first demonstrations of
the power of broadcast radio, and its rapid commodification into corporate capital via the new "durative form" of the phonographic recording was something new. The capitalist productive matrix plays a
central role in this adrenalinized technonarrative process: as Murray
and Brucker point out (1979), the intense competition between the
two Louisville papers and a relatively slow period in the news provided an economic incentive to exploit the events at Sand Cave to

their limits. The effect of such competition puts particular pressure on the binary that Schudson calls "the two journalisms": one stressing the rational ideal of information, the other fulfilling an older storytelling function (1978, 88). It happens that just two years after the Sand Cave Tragedy, sociologist Robert Park would point out that the format of news and of fiction had become nearly indistinguishable (cited in Gabler 1998, 74). Kafka's "fictional" narrative was based, as B. Mitchell convincingly demonstrates, on the "real" commercial spectacle of circus hunger artists, while the term "tragedy" in the "real" narrative of the Sand Cave *Tragedy* signals an interest in the "fictional." The capitalist technoeconomic system exploits the event for all it's worth, from the initial fanfare of Carnival Sunday (where again, a kind of semiotic clutter obscures the "real" event) and the driven, competitive, industrial pace of the news-reporting systems, to hit records, a Hollywood movie, and today, a tourist attraction at Sand Cave, Kentucky, and a musical score on compact disk. Again, we see that the Sand Cave Tragedy was "mined" for all its stored resources in terms of the capital to be derived from entertainment. This "mining," we might add, is a process we saw very clearly in a number of places. We saw it in the transference of rap from a vocal form to an element within Hollywood narratives (the urban resource of rap mined and transformed). We also saw it in nostalgic entertainment, in which the recent past is mined for profits, thus triggering a "foreshortening" of nostalgia as that past which is being mined comes dangerously close to the terra firma of the present and we have a kind of marketed nostalgia for cultural products that are only a few years old. Such mining is likewise characteristic of the pornography industry, which constantly digs into an ever-renewing mine of female bodies to keep the machine going.

Another popular-modern feature of the Collins story is the shifting of focus from the alleged event to the secondary technical apparatus that initially serves the event. Collins and his heroic isolation were the alleged focus of the event, but sometimes it is hard to tell if the real hero of the story is Collins or the Louisville journalist 'Skeets' Miller. If this be the case, the older myth of the hero's journey (and Collins does attempt a journey through an underworld, thus conforming to Campbell's myth) has been supplanted by two modern myths: first, that of the industrial and technological power of modern society (exemplified by the earth-moving machinery and the "know-how" of the mining engineers); and second and relatedly, that of the media hero, who here is a focal point for a celebration of the ideals of an industrial democracy (he energetically "digs into" the story, mining it for its "truth"). Here we see that the secondary technology not only conveys the mythic content: it becomes thematized. In summary, then, an actual (historical) event (man caught in a cave, ensuing rescue attempt), becomes mythohistoricized, becomes a myth of figurative "man" in mythic battle with the underworld, here presented in the Christian version, with Dalhart's "Ballad of Floyd Collins" (warning us to "get right with [our] maker / Before it's too late" and the Sunday sermons rendering the Collins affair typologically, as "sin-cursed man trapped by the rock of evil." Additionally, the Sand Cave Tragedy conveys a newer myth, that of industrial man, whose mastery of technology and capital grant him at least the possibility of overcoming nature. And finally, the Collins case has particular interest for us as a story that characterizes popular modernity. The Collins event, swallowed up by a system of information, and ultimately entertainment, is the story of commodified man—a man turned into a product.

Watching wrestling in its earlier, authentic form, in which the athletes were often locked in stationary positions for long stretches of time, was, I would imagine, something like watching the hunger artist at his fast. The reorganization of the sport in the 1920s and ongoing reorganization movements in every decade since the 1950s were aimed at adding long-term interest (Jenkins details the economic rationale here [1997, 49]) in the form of grudge and feud narratives, thus giving the event itself more diachronic appeal. These changes have the same effect as the impresario's forty-day limit and his addition of theatrical flourishes for the hunger artist's emergence from his cage. Like the impresario, the wrestler, as Barthes says, exploits the spectacle of suffering "to the limits of [its] meaning" ([1957] 1977, 16). We see, then, that wrestling became transformed; it went from athletic competition to stage-managed spectacle and, finally, to a masculinist serial narrative (Jenkins 1997).

We have seen that professional wrestling was largely shaped by new media: first, by its failure with radio, and then, its great success with television in the 1950s and the new cable networks in the 1980s. Indeed, contemporary professional wrestling is a massively hypermediated spectacle in which the staged spectacle in the ring not only spills out into the aisles but spills out semiotically into the commentators' room with their elaborate video monitor displays, into the comments of other wrestlers backstage captured on remote camera, into off-site locations. This hypermediation effect is an integral part of the stage-managed spectacle of professional wrestling, which, like the porno-graphic spectacle, is an "elaborate construction" for the benefit of the viewer—"they perform to be observed" (Kimmel 1990a, 311). Thus, for the viewer at home, the match is mediated by the screen, mediated by the nondiagetic discourse of the commentators, and mediated by the narratives that occur between matches—statements by the wres-tlers, "analyses" by commentators, often posed in control rooms with multiple television monitors, which thus visually embody the web of mediation. This scene implies a particular approach to reality and the production of meaning, embodying Debord's claim that the spectacle manifests itself as a "weltanschauung that has been actualized into the material world" ([1967] 1983, 13, sec. 5).

In the Reagan presidency we see the same kind of managerial apparatus at work, which no doubt played a role in the American presidency as far back as the "Virginia Dynasty," but began in earnest with Andrew Jackson. In terms of modern media technology, let us consider for one final time FDR's Fireside Chats. It is worth noting that while Ryan calls Roosevelt's presidency a "rhetorical" one, he really implies that it was more technorhetorical in nature, for as he says, FDR credited "radio and the moving picture, or the newsreels, and not the newspapers, for having helped inform the people about the needs of their government" (1988, 30). With FDR, then, the symbiotic relationship between electronic media and the older print journalism—a relationship established in the 1844 Baltimore-to-Wash-ington telegraph transmission of the Whig nominating convention results and the quick reaction of the press in terms of disseminating that information—becomes inverted, with the electronic media be-coming an instrument of political will and the major source of infor-mation for the masses.

Ronald Reagan's presidency certainly bore the stamp of FDR's "rhetorical presidency," and if we take a view of history that gives priority to communications theory rather than political theory, we see, quite paradoxically, that while Reagan advocated the unraveling of the form of federal government established by FDR, in another respect he was truly FDR's protégé. Reagan developed his savvy with radio in the mid-1930s, and from his screen career he acquired a similar savvy with video. His presidency is thus a continuation of the technorhetorical modality of FDR. Gary Wills would agree, for he argues that the automobile, the radio, and the television form the technological triumvirate that was absolutely essential to the Reagan phenomenon (1988, 372–73), and furthermore, that Reagan's significance is not only political but technological, for he was, as Wills perceptively notes, a living "link between the pioneering era of broadcast technology and the sophisticated politics of our modern communications industry" (130). As such, Reagan's genius was to recognize, to an unprecedented degree, the Habermasian recognition of the "indissoluble link between the institutions and practices of mass public communication and the institutions and practices of democratic politics" (Garnham 1992, 360). Reagan's knowledge of the media and American mythology, formidable enough on its own, was enhanced by the knowledge of his campaign managers and White House aides. Needless to say, the elements of presidential campaigns are calculated to exploit mythohistoric resonances to their very limit. The previously mentioned visits to the former Reagan home in Dixon, Illinois, were used, for example, to create an updated version of "one-room log cabin" mythos of Abraham Lincoln, to say nothing of the manger in Bethlehem. These factors no doubt compelled Edmund Morris, in his *Dutch: A Memoir of Ronald Reagan* (1999) to approach his subject using the methods of fiction (i.e., using a fictional narrator) as much as those of traditional (i.e., "factual") biography.

Popular modernity's narrative-technology nexus, then, is found in the general conflation of fictional (mythical, prerationalist) narrative, nonfictional (empirical, rationalist) exposition, and the secondary distribution systems that deliver these narratives and expositions. In popular modernity, these elements, which were always closely related, are reduced to a nearly indistinguishable tangle. In the case of Floyd Collins, we have noted the efforts of journalists to emphasize the dramatic virtues of the event. The fact that it was the Sand Cave *Tragedy* rather than a mere incident as well as its transformation into a pseudofolk ballad is telling enough on this point. The 1980s, the Age of Reagan, were particularly productive in this regard: it is worth considering that during the mid-1980s, the *Weekly World News*, a parody of the *National Enquirer*, was one of America's fasting growing newspapers. Similarly, with its reemergence as a spectacular entertainment, professional wrestling was, with basketball, one of the fastest growing spectator "sports," or rather, "sports entertainment," to use the Worldwide Wrestling Federation's official term—a term that reveals the popular modern concatenation of reality (i.e., "sports," which connotes "real" competition) and fantasy ("entertainment," in which being amused takes precedence over authenticity). "The public," Barthes says of the wrestling event, "is completely uninterested in knowing whether the contest was rigged or not, and rightly so; it abandons itself to the primary virtue of the spectacle, which is to abolish all motives and consequences; what matters is not what it thinks but what it sees" ([1957] 1977, 15). The fact that a sport that isn't really a sport and a newspaper that isn't really a newspaper should reach such heights during the era of the actor/president is indeed suggestive, and Reagan's career certainly exemplifies this aspect of popular modernity. The 1998 election to the governorship of Minnesota of former professional wrestler Jessie "the Body" Ventura, whose public persona and fame was more than a match for both Republicans and Democrats, only serves to confirm these observations.

Ronald Reagan's first job in entertainment was creating "dramatic" accounts of baseball games based on the barest outlines supplied over a wire, and we have noted that once when the wire failed, Reagan continued to simply "make up" the game. This was the beginning of what Gary Wills called Reagan's "complicity in make believe" (1988, 119). We noted earlier how John Hinckley's failed attempt on President Reagan's life served to bolster Reagan's mythic status. We must add that in this Hinckley imitated the fictional Travis Bickle (the mentally disturbed would-be assassin of Martin Scorsese's film, *Taxi Driver* [1976]). Thus in a strange way Hinckley is Reagan's *Doppelgänger*, for Reagan's appeal was in turn an imitation of Hollywood film; his main source for slogans was the words of action-adventure heros—Stallone, Schwartzenegger, and most memorably, Clint Eastwood's character Dirty Harry, the source for Reagan's quip, "Go ahead—make my day." In the epoch of popular modernity, it would seem that both the legitimate mainstream politician and the isolated sociopath are informed by the same hyperproductive narrative matrix.

Reagan's Hollywood connection goes well beyond the occasional borrowed line from the movies. For instance, in *Tropic Zone* (1953), Reagan plays Dan McCloud, a North American who takes a job as foreman of a Central American banana plantation. McCloud learns that his boss, an evil foreigner named Lukats, is plotting to monopolize the banana exporting trade. McCloud—with the assistance of one of Lukats's smaller competitors, a North American woman—organizes the workers and they manage to prevent Lukats's corrupt machinations. The U.S. interference in third-world politics in the post–World War II era is unmistakably reflected in this B-grade film, in which there is the transference of a standard plot of Westerns: this film, says Thomas, is "actually a western, with plantations instead of ranches and bananas instead of cattle" (1980, 199). *Tropic Zone* may strike one as prophetic, given Reagan's Central American policy as it was conducted through his surrogate, Oliver North (who reversed his employer's career pattern—whereas Reagan started in radio and ended up in government, North started in government and ended up in radio, starting a career as a conservative talk show personality in 1995).

In his study of the dominant role entertainment plays in all of public life, Gabler notes the difference between Reagan and predecessors like Kennedy and Nixon (both skilled technorhetoricians). He says that Reagan "had so thoroughly internalized the cosmology of the movies that he . . . lived entirely within it" (1998, 110). When we watch other movies from Reagan's film career, we find pervasive evidence of this. A good example of this is *Murder in the Air* (1940), in which, according to the movie poster, the "Secret Service battle for the secret of the most terrifying weapon ever invented! . . . a mystery gun . . . and its ray of horror!" This screenplay, Rogin notes, gave Reagan the idea for the Strategic Defense Initiative that he attempted to enact during his presidency, and SDI itself was named after a movie, *Star Wars* (1998, 23). In any case, all of Reagan's adventure films, whether set in Central America or in some technospy scenario, refer back to that standard Western plot—the lawman who "cleans up the town," an action paralleled by Reagan's "laying down the law" with the Air Traffic Controllers Union and government social programs. During his film career Reagan "had a persistent desire to make Westerns," a habit he may have carried over into his political career (Thomas 1980, 199). In so doing, Reagan successfully connected, through the adroit use of media technologies, with the American mythos of individualism and America's nostalgia for a mythic individualist past as represented by the figure of the lone frontiersman. This in spite of the fact that, as Wills notes, the frontier was settled, not by lone heroes, but by groups and institutions—by federal troops, railroad companies, and civic organizations (1988, 380). The phenomenon of Reagan's popularity during the 1980s—a popularity, as studies reveal, that was based largely on the appeal of his personality—not only reveals the American electorate's "need for illusion" (Alford 1988, 581, 572): it also reveals the complicity of the American political and entertainment systems. It is also telling that Phil Dusenberry, BBD&O creative wunderkind, was a major figure in forming, first, the Pepsi advertising campaigns of the 1960s and 1970s; second, a popular, nostalgia-based "baseball" film based on a Bernard Malamud novel, *The Natural* (1984); and third, the Reagan-Bush presidential campaigns (Smithsonian files). The fact that the same person could be involved in the shaping of three purportedly different kinds of activity (film narrative, political campaign, product advertising) is explained by Baudrillard in a statement that grasps one of the essentials of popular modernity. "Governing today," he says, ". . . is like advertising and it is the same effect that is achieved—commitment to a scenario, whether it be a political or an advertising scenario" (qtd. in Nadel 1997, 22).

I have been talking about narratives in this chapter, and this book, itself constituting in some sense a narrative, is now finished. In closing, let me first acknowledge the limitations of all the above. First, cultural criticism is implicated in the mythologies it critiques. Second, the act of periodization in cultural history is problematized by transhistorical elements, such as myth, as well as the problem of trying to reconstruct subjective experience. Third, the analytical terms employed throughout—"experience," "technology," "mythohistory"—are less separate entities than overlapping spheres or Chinese boxes that collapse one into the other. Nevertheless, I hope I've accomplished something of my task here, which has been to demonstrate with some measure of warranted assertability the dynamics and specific modality of popular modernity in America.

Works Consulted

Ace in the Hole. 1950. Film. Directed by Billy Wilder. Screenplay by Billy Wilder. Paramount. Re-released as *The Big Carnival* (Paramount, 1950).

Ackroyd, Peter. 1984. *T. S. Eliot: A Life.* New York: Simon & Schuster.

Adorno, Theodor. [1951] 1978. *Minima Moralia.* Trans. E. F. N. Jephcott. London: Verso.

Adorno, Theodor, and Max Horkheimer. [1944] 1993. "The Culture Industry." In *The Cultural Studies Reader,* ed. Simon During, 30–43. London: Routledge.

Alford, C. Fred. 1988. "Mastery and Retreat: Psychological Sources of the Appeal of Ronald Reagan." *Political Psychology* 9, no. 4:571–89.

Alkon, Paul K. 1994. *Science Fiction Before 1900.* New York: Twayne.

Allen, Ernest J. 1996. "Making the Strong Survive: The Contours and Contradictions of Message Rap." In Perkins 1996b, 159–91.

Allman, Greg. 1995. Interview. In *Up from the Underground,* vol. 11 of *The History of Rock and Roll.* Time/Life Videotape.

Alter, Robert. 1993. "Modernism and Nostalgia." *Partisan Review* 60, no. 3:388–403.

Altman, Rick. 1987. *The American Film Musical.* Bloomington: Indiana University Press.

American Graffiti. 1973. Film. Directed by George Lucas. Screenplay by George Lucas and Gloria Katz. Universal.

Anders, Peter. 1994. "The Architecture of Cyberspace." *Progressive Architecture* 75, no. 10:78–82, 101.

Andersen, Christopher. 1993. *Jagger Unauthorized.* New York: Delacorte.

Anderson, Benedict. [1991] 1994. "Imagined Communities." In Hutchinson and Smith, 89–95.

Apollo 13. 1995. Film. Directed by Ron Howard. Screenplay by William Broyles Jr. and Al Reinert. Universal.

Anzieu, Dider. 1989. *The Skin Ego.* Trans. Chris Turner. New Haven: Yale University Press.

Aronson, S. 1971. "The Sociology of the Telephone." *International Journal of Comparative Sociology,* 12 September, 153–67.

Asante, Molifi. 1987. *The Afrocentric Idea.* Philadelphia: Temple University Press.

Attali, Jacques. 1989. *Noise.* Minneapolis: University of Minnesota Press.

Auerbach, Nina. 1982. *Woman and the Demon.* Cambridge: Harvard University Press.

Bachelard, Gaston. 1969. *The Poetics of Space.* Boston: Beacon.

Back to the Future. 1985. Film. Directed by Robert Zemeckis. Screenplay by R. Zemekis and Bob Gale. Produced by Steven Spielberg. Universal.

Back to the Future, Part II. 1989. Film. Directed by Robert Zemeckis. Produced by Steven Spielberg. Universal.

Back to the Future, III. 1990. Film. Directed by Robert Zemeckis. Screenplay by Robert Zemeckis and Bob Gale. Universal.

Baker, Houston A. 1993. *Black Studies, Rap, and the Academy.* University of Chicago Press.

Baker, Patrick L. 1993. "Space, Time, Space-Time and Society." *Sociological Inquiry* 63, no. 4 (November): 406–24.

Bakhtin, M. M. 1981. *The Dialogic Imagination.* Austin: University of Texas Press.

Baldwin, James. 1990. "Sonny's Blues." In *The Norton Anthology of Short Fiction,* 4th ed., ed. R. V. Cassill, 23–50. New York: Norton.

Ball, Donald W. 1968. "Towards a Sociology of Telephones." In *Sociology and Everyday Life,* ed. Marcello Truzzi, 59–74. Englewood Cliffs, N.J.: Prentice-Hall.

Bambara, Toni Cade. 1993. "Reading the Signs, Empowering the Eye: *Daughters of the Dust* and the Black Independent Cinema Movement." In Diawara 1993b, 118–44.

Baraka, Amiri. 1993. "Spike Lee at the Movies." In Diawara 1993b, 145–53.

Barlow, William. 1990a. "Cashing In: 1900–1939." In Dates and Barlow 1990b, 25–56.

———. 1990b. "Commercial and Noncommercial Radio." In Dates and Barlow 1990b, 175–250.

Barnouw, Eric. 1975. *Tube of Plenty: The Evolution of American Television.* New York: Oxford University Press.

Barthes, Roland. [1957] 1977. *Mythologies.* New York: Hill and Wang.

———. 1980. "Upon Leaving a Movie Theatre." In *Apparatus,* ed. T. H. Kyung. New York: Tanam Press.

———. 1981. *Camera Lucida: Reflections on Photography.* Trans. Richard Howard. New York: Hill and Wang.

Bataille, George. 1977. *Death and Sensuality*. New York: Arno. Trans. of *L'Erotisme* (Paris: Editions de Minuit, 1957).

————. 1979. "The Psychological Structure of Fascism." *New German Critique* 16 (winter): 64–87.

Bateson, M. C. 1990. *Composing a Life*. New York: Plume.

Baudrillard, Jean. 1994. *Simulacra and Simulation*. Trans. Sheila Faria. Ann Arbor: University of Michigan Press.

Bayner, Norman H. 1969. *The Speeches of Adolph Hitler*. New York: Howard Fertig.

Bayton, Mavis. 1997. "Women and the Electric Guitar." In *Sexing the Groove: Popular Music and Gender,* ed. Sheila Whitely, 37–49. London and New York: Routledge.

Beauvoir, Simone de. 1966. "Must We Burn Sade?" Introduction to *The One Hundred and Twenty Days of Sodom and Other Writings,* by the marquis de Sade, comp. and trans. Austryn Wainhouse and Richard Seaver, 3–64. London: Arrow Books.

Belk, Russell W. 1991. "Possessions and the Sense of the Past." In *Highways and Buyways,* ed. Association for Consumer Research, 114–30. Provo, Utah: Association for Consumer Research.

Bell, Alexander Graham. 1908. *The Bell Telephone: The Deposition of Alexander Graham Bell . . .* Boston: Bell Telephone.

Bell, Ella. 1990. "The Bicultural Life Experiences of Career-Oriented Black Women." *Journal of Organizational Behavior* 11, no. 6:459–77.

Beneke, Timothy. 1990. "Intrusive Images and Subjectified Bodies: Notes on Visual Heterosexual Porn." In Kimmel 1990b, 168–89.

Benjamin, Walter. 1973. *Illuminations*. London: Fontana.

Bennett, Tony. 1983. "The Bond Phenomenon: Theorizing a Popular Hero." *Southern Review: Literary and Interdisciplinary Essays* 16, no. 2 (July): 195–225.

Bennett, Tony. 1989. Interview. In *Satchmo Louis Armstrong*. Videotape. Directed by Gary Giddens, Kendrick Simmons, Hattie Winston, and Melvin Van Peebles. Masters of American Music Series. CBS. Based on the book by Gary Giddins.

Benshoff, Harry M. 1997. *Monsters in the Closet: Homosexuality and the Horror Film*. Manchester: Manchester University Press.

Bercovitch, Sacvan. 1978. *The American Jeremiad*. Madison: University of Wisconsin Press.

Berger, Bennett. 1970. "Black Culture or Lower-Class Culture?" In *Soul,* ed. Lee Rainwater, 131–42. Chicago: Aldine.

Berger, John. 1972. *Ways of Seeing.* London: BBC/Penguin.

Berland, Jody. 1992. "Angels Dancing: Cultural Technologies and the Production of Space." In *Cultural Studies,* ed. Lawrence Grossberg and Nelson Cary, 38–51. New York: Routledge.

Bernardini, Gene. 1982. "Anti-Semitism." In Cannistraro 1982b, 28–31.

Bernasconi, Robert. 1993. "On Deconstructing Nostalgia for Community within the West: The Debate between Nancy and Blanchot." *Research in Phenomenology* 23 (fall): 3–21.

Birkerts, Sven. 1989. "The Nostalgia Disease." *Tikkun* 4, no. 2 (March/ April): 20–22+.

Black, Joel. 1991. *The Aesthetics of Murder.* Baltimore: Johns Hopkins University Press.

Blanchard, Marc. 1992. "His Master's Voice." *Studies in the Literary Imagination.* 1, no. 25 (spring): 61–79.

Blauner, Robert. 1970. "Black Culture: Myth or Reality?" In *Afro-American Anthropology,* ed. Norman E. Witten, 347–66. New York: Free Press.

Blaut, J. M. 1993. *The Colonizer's Model of the World: Geographical Diffusionism and Eurocentric History.* New York: Guilford Press.

Blondheim, Menahem. 1994. *News over the Wires: The Telegraph and the Flow of Public Information in America, 1844–1897.* Cambridge: Harvard University Press.

Bly, Robert. 1992. *Iron John: A Book About Men.* New York: Vintage.

Bodkin, Joseph. 1986. *Sambo: The Rise and Demise of an American Jester.* New York: Oxford University Press.

Bordo, Susan. 1993. "Reading the Male Body." *Michigan Quarterly Review* 32, no. 4 (fall): 696–737.

Boulding, Kenneth E. 1956. *The Image.* Ann Arbor: University of Michigan Press.

Boyarin, Jonathan, ed. 1994. *Remapping Memory: The Politics of TimeSpace.* Minneapolis: University of Minnesota Press.

Boyd, Todd. 1997. *Am I Black Enough for You? Popular Culture from the 'Hood and Beyond.* Bloomington: Indiana University Press.

Bradford, William. 1952. *Of Plymouth Plantation.* Ed. Samuel Eliot Morison. New York: Knopf.

The Brady Bunch Movie. 1995. Film. Directed by Betty Thomas. Script by Sherwood Schwartz and Laurice Elehwany. Paramount.

Brennan, Timothy. 1993. "The National Longing for Form." In *Nation and Narration,* ed. Homi K. Bhabha, 44–70. New York: Routledge.

Brigadoon. 1954. Film. Directed by Vincenti Minnelli. Screenplay by. Alan Jay Lerner. MGM.

Brigham, Albert Perry. 1921. "Geographic Education in America." In *Smithsonian Institution Annual Report, 1919*, 487–97. Washington, D.C.: Smithsonian Institution.

Brigham, Albert Perry, and Richard E. Dodge. 1933. "Nineteenth-Century Textbooks of Geography." In *The Thirty-Second Yearbook of the National Society for the Study of Education: The Teaching of Geography*, ed. Guy Montrose Whipple, 3–27. Bloomington, Ill.: Public School Publishing.

Brown, Robert J. 1998. *Manipulating the Ether: The Power of Broadcast Radio in Thirties America*. Jefferson, N.C.: McFarland.

Brown, Sterling. 1937. *The Negro in American Fiction*. New York: Arno Press.

Browne, Ray, and Ronald J. Ambrosetti, eds. 1993. *Continuities in Popular Culture*. Bowling Green, Ohio: Bowling Green State University Press.

Bruck, Connie. 1997. "The Takedown of Tupac." *The New Yorker*, 7 July, 46–64.

Bruner, J. 1990. *Acts of Meaning*. Cambridge: Harvard University Press.

Burke, Peter. 1978. *Popular Culture in Early Modern Europe*. New York: New York University Press.

———. 1980. *Sociology and History*. London: Allen & Unwin.

Burns, Gary. 1994. "How Music Video Has Changed and How It Has Not Changed." *Popular Music & Society* 18, no. 3 (fall): 67–79.

Caldwell, John Thornton. 1995. *Televisuality: Style, Crisis, and Authority in American Television*. New Brunswick, N.J.: Rutgers University Press.

Calkins, Ernest Elmo. 1928. *Business, the Civilizer*. Boston: Little, Brown.

Cannistraro, Philip V. 1972. "The Radio in Fascist Italy." *Journal of European Studies* 2:127–54.

———. 1982a. "Radio." In Cannistraro 1982b, 446–47.

———, ed. 1982b. *Historical Dictionary of Fascist Italy*. Westport, Conn.: Greenwood Press.

Cantor, Norman. 1988. *Twentieth-Century Culture*. New York: Peter Lang.

Cantril, Hadley. 1940. *The Invasion from Mars: A Study in the Psychology of Panic*. Princeton: Princeton University Press.

Cantril, Hadley, and Gordon W. Allport. [1935] 1986. *The Psychology of Radio*. New York: Harper. Reprint, Salem, N.H.: Ayer.

Carey, James. 1991. "Time, Space, and the Telegraph." In Crowley and Heyer 1991, 132–37.

Carpenter, Edmund, and Marshall McLuhan. 1960. "Acoustic Space." In *Explorations in Communication*, ed. Carpenter and McLuhan. Boston: Beacon Press.

Carpenter, Humphrey. 1977. *Tolkien: A Biography*. Boston: Houghton Mifflin.

Carpenter, Lucas. 1995. "Floyd Collins and the Sand Cave Tragedy: A Possible Source for Faulkner's *As I Lay Dying*." *The Kentucky Review* 12, no. 3 (spring): 3–18.

Carr, David. 1986. *Time, Narrative, and History*. Bloomington: Indiana University Press.

Carroll, David. 1995. *French Literary Fascism*. Princeton: Princeton University Press.

Carroll, Michael P. 1989. *Catholic Cults and Devotions: A Psychological Inquiry*. Kingston, Montreal, and London: McGill-Queen's University Press.

Carroll, Michael T. 1993. "The Bloody Spectacle: Mishima, The Sacred Heart, Hogarth, Cronenberg, and the Entrails of Culture." *Studies in Popular Culture* 15, no. 2:43–56.

Carroll, Noël. 1990. *The Philosophy of Horror; or, Paradoxes of the Heart*. New York: Routledge.

Carter, Steven. 1994. "On American Time: Mythopoesis and the Marketplace." *Journal of American Culture* 17, no. 2 (summer): 35–40.

CB4. 1993. Film. Directed by Tamra Davis. Screenplay by Chris Rock, Nelson George, and Robert Locash. Produced by Nelson George. MCA/Universal.

Class Act. 1992. Film. Directed by Randall Miller. Screenplay by Michael Swerdlick, Wayne Rice, and Richard Brenne. Warner Brothers.

Cloke, Paul, Chris Philo, and David Sadder. 1991. *Approaching Human Geography*. New York: Guilford Press.

Clover, Carol. 1991. *Men, Women and Chain Saws: Gender in the Modern Horror Film*. Princeton: Princeton University Press.

Collins, Margery L., and Christine Pierce. 1976. "Holes and Slime in Sartre." In *Woman and Philosophy*, ed. Carol Gold and Marx Wartofsky, 112–27. New York: Putnam.

Colunga. Jeannie Marie. 1993. "We Have Nothing to Fear but Tropes Themselves: Rhetoric in the Speeches of Franklin Delano Roosevelt." M.A. thesis. California State University at San Bernardino.

Combs, James. 1993. *The Reagan Range: The Nostalgic Myth in American Politics*. Bowling Green, Ohio: Popular Press.

Coontz, Stephanie. 1992. *The Way We Never Were: American Families and the Nostalgia Trap*. New York: Basic Books.

Cornford, F. M. 1936. "The Invention of Space." In *Essays in Honor of Gilbert Murray*, ed. J. A. K. Thomson and Arnold Joseph Toynbee, 215–35. London: Allen & Unwin.

"Cream Pie: Your Source for Internal Cum Shots." 1999. WWW.creampie. com.

Cripps, Thomas. 1975. *Slow Fade to Black.* New York: Oxford University Press.

———. 1979. *Black Film as Genre.* Bloomington: Indiana University Press.

———. 1990. "Making Movies Black." In Dates and Barlow 1990b, 125–74.

Crisell, Andrew. 1986. *Understanding Radio.* London and New York: Routledge.

Crosby, Bing. 1989. *Pocketful of Dreams.* American Legends Series. Fanfare CDD 457 Compact Disc. Digitally restored recordings of selected Crosby recordings from 1931 to 1938.

Crowley, David, and Paul Heyer, eds. 1991. *Communication in History.* New York: Longman.

Csikszentmihalyi, Mihaly, and Eugene Rochberg-Halton. 1978. "People and Things: Reflections on Materialism." *The University of Chicago Magazine,* spring, 7–8.

Czitrom, Daniel J. 1982. *Media and the American Mind.* Chapel Hill: University of North Carolina Press.

Dates, Hannette L., and William Barlow. 1990a. "A War of Images." In Dates and Barlow 1990b, 1–24.

———, eds. 1990b. *Split Image: African Americans in the Mass Media.* Washington, D.C.: Howard University Press.

Davis, Angela Y. 1994. "Afro Images: Politics, Fashion, and Nostalgia." *Critical Inquiry* 21 (autumn): 37–45.

Davis, Fred. 1979. *Yearning for Yesterday.* New York: Free Press.

Davis, Steven. 1985. *Hammer of the Gods: The Led Zeppelin Saga.* New York: Ballantine.

Debord, Guy. [1967] 1983. *Society of the Spectacle.* Detroit, Mich.: Black and Red.

De Camp, L. Sprague. 1975. *H. P. Lovecraft: A Biography.* New York: Barnes and Noble.

D'Emilio, John, and Estelle B. Freedman. 1988. *Intimate Matters: A History of Sexuality in America.* New York: Harper.

Derrida, Jacques. 1973. *Speech and Phenomena.* Evanston, Ill.: Northwestern University Press.

Deutsch, Karl. 1953. *Nationalism and Social Communication.* New York: Wiley.

———. 1963. *The Nerves of Government.* New York: Free Press.

Diawara, Manthia. 1993a. "Black American Cinema: The New Realism." In Diawara 1993b, 3–25.

———, ed. 1993b. *Black American Cinema.* New York: Routledge.

Dick, Bernard F. 1996. *Billy Wilder*. Rev. ed. New York: Da Capo.

Di Lauro, Al. 1976. *Dirty Movies: An Illustrated History of the Stag Film*. New York: Chelsea House.

Dos Passos, John. 1934a. *In All Countries*. New York: Harcourt, Brace.

———. 1934b. "The Radio Voice" *Common Sense*, February, 17.

Douglas, Susan J. 1987. *Inventing American Broadcasting, 1899–1922*. Baltimore: Johns Hopkins University Press.

———. 1991. "Broadcasting Begins." In Crowley and Heyer 1991, 190–97.

Du Bois, W. E. B. 1903. *The Souls of Black Folk*. Chicago: McClurg.

du Moncel, Th. comte. 1880. *The Telephone, the Microphone, and the Phonograph*. London: Kegan Paul. First published in French by Hachette Press (Paris, 1878).

Dunn, Leslie C., and Nancy Jones. 1994a. Introduction to Dunn and Jones 1994b, 1–13.

———, eds. 1994b. *Embodied Voices: Representing Female Vocality in Western Culture*. Cambridge and New York: Cambridge University Press.

Dyson, Michael Eric. 1992. "Out of the Ghetto." *Sight and Sound* 2, no. 6 (October): 18–21.

———. 1993. *Reflecting Black: African-American Cultural Criticism*. Minneapolis: University of Minnesota Press.

———. 1996. *Between God and Gangsta Rap*. New York: Oxford University Press.

Eatwell, Roger. 1966. *Fascism: A History*. New York: Penguin.

Edmands, B. Franklin. 1832. *The Boston School Atlas, Embracing a Compendium of Geography*. Boston: Lincoln & Edmands.

Elazar, Daniel J. 1994. *The American Mosaic: The Impact of Space, Time, and Culture on American Politics*. Boulder, Colo.: Westview.

Eliot, T. S. 1934. *After Strange Gods*. New York: Harcourt Brace.

Ellison, Ralph. 1947. *Invisible Man*. New York: Vintage.

———. 1995. Interview, with Hollie I. West. "Ellison: Exploring the Life of a Not So Visible Man." In *Conversations with Ralph Ellison*, ed. Maryemma Graham and Amritjit Singh, 235–58. Jackson: University of Mississippi Press.

Ellul, Jacques. 1970. *The Technological Society*. New York: Knopf.

Elmer-DeWitt, Philip. 1995. "Welcome to Cyberspace." *Time*. 145, no. 12 (spring): 4–11. Special issue.

Emerson, Ken. 1997. *Doo-dah! Stephen Foster and the Rise of American Popular Culture*. New York: Simon & Schuster.

Emerson, Rupert. 1962. *From Empire to Nation.* Cambridge: Harvard University Press.

Ewen, Stuart. 1976. *Captains of Consciousness: Advertising and the Social Roots of the Consumer Culture.* New York: McGraw-Hill.

———. 1988. *All-Consuming Images: The Politics of Style in Contemporary Culture.* New York: Basic Books.

Fear of a Black Hat. 1993. Film. Directed by Rusty Cundieff. Screenplay by Rusty Cundieff. 1993. ITC Entertainment.

Ferden, Paul. 1952. "The Ego as a Subject and Object in Narcissism." In *Ego Psychology and the Psychoses,* ed. E. Weiss, 283–322. New York: Basic Books.

Ferguson, Andrew. 1994. "Stayin' Alive." *National Review* 46, no. 2 (7 February): 80.

Feuer, Jane. 1993. *The Hollywood Musical.* 2d ed. Bloomington: Indiana University Press.

Fischer, Claude S. 1992. *America Calling: A Social History of the Telephone to 1940.* Berkeley: University of California Press.

Fiske, John. 1991. *Understanding Popular Culture.* New York: Routledge.

Flanagan, Owen. 1992. *Consciousness Reconsidered.* Cambridge: MIT Press.

Forbes, Jack D. 1968. "Frontiers in American History and the Role of the American Historian." *Ethnohistory* 15 (spring): 203–35.

Foucault, Michel. 1986. "Of Other Spaces." *Diacritics,* spring, 22–27.

Frazier, E. Franklin. 1957. *Black Bourgeoisie.* London: Collier-Macmillan.

Fresh Prince of Bel Air. 1990. Starring Will Smith. NBC series.

Friday. 1995. Film. Directed by F. Gary Grey. Produced by Ice Cube. Screenplay by Ice Cube and D. J. Pooh. New Line.

Frith, Simon. 1986. "Art Versus Technology: The Strange Case of Popular Music." *Media, Culture, and Society* 8:259–78.

———. 1996. "Music and Identity." In *Questions of Cultural Identity,* ed. Stuart Hall and Paul du Gay, 108–27. London: Sage.

Fromm, Erich. 1941. *Escape From Freedom.* New York: Farrar and Rinehart.

———. 1947. *Man for Himself.* New York: Rinehart.

Frye, Northrop. 1957. *Anatomy of Criticism.* Princeton: Princeton University Press.

Fukuyama, Francis. 1989. "The End of History?" Supplement to *The National Interest,* summer, 2+.

Fulwood, Sam. 1991. "The Rage of the Black Middle Class." *Los Angeles Times Magazine,* 3 November, 22–24+.

Fussell, Sam. 1993. "Bodybuilder Americanus." *Michigan Quarterly Review* 32, no. 4 (fall): 576–96.

Gabbard, Glen O., and Eva P. Lester. 1995. *Boundaries and Boundary Violations in Psychoanalysis.* New York: Basic Books.

Gabler, Neal. 1998. *Life, the Movie: How Entertainment Conquered Reality.* New York: Knopf.

Garafalo, Reebee. 1990. "Crossing Over: 1939-1989." In Dates and Barlow 1990b, 57–124.

Garnham, Nicholas. 1992. "The Media and the Public Sphere." In *Habermas and the Public Sphere,* ed. Craig Calhoun, 359–76. Cambridge: MIT Press.

George, Nelson. 1994. *Blackface: Reflections on African Americans and the Movies.* New York: HarperCollins.

"Germany Acts to Ban Songs by Five Neo-Nazi Rock Groups." 1992. *The New York Times,* 3 December, A12 col. 1.

Gibson, William. 1987. *Count Zero.* New York: Ace Books.

Gilmore, Paul. 1996. "The Telegraph in Black and White." Paper presented at the forum Comparative Technologies: The Telegraph and the Internet, MLA Convention, Washington D.C., December.

Goodwin, Andrew. 1987. "Music Video in the (Post) Modern World." *Screen* 28:36–55.

Graham, Allison. 1984. "History, Nostalgia, and the Criminality of Popular Culture." *The Georgia Review* 38, no. 2:348–64.

Grant, Barry Keith, ed. 1996. *The Dread of Difference: Gender and the Horror Film.* New York: Manchester University Press.

Graves, N. 1975. *Geography in Education.* London: Heinemann.

Griffin, Susan. 1981. *Pornography and Silence.* New York: Harper.

Gubar, Susan, and Joan Hoff, eds. 1989. *For Adult Users Only: The Dilemma of Violent Pornography.* Bloomington: Indiana University Press.

Guns 'N' Roses. 1988. *G 'N' R Lies.* Compact disk. UNI/Geffen.

Habermas, Jürgen. 1989. "The Public Sphere." In *Jürgen Habermas on Society and Politics.* Boston: Beacon. Originally published as "Öffentlichkeit," in *Kultur und Kritik,* by Jürgen Habermas (Frankfurt: Suhrkamp Verlag, 1973).

Haden-Guest, Anthony. 1997. "World's Biggest Gang Bang II." *Penthouse* 28, no. 10 (June): 134–36+.

Halberstam, Judith. 1993. "Technologies of Monstrosity: Bram Stoker's *Dracula.*" *Victorian Studies,* spring, 332–52.

Hall, G. Stanley. 1911. *Educational Problems.* New York and London: Appleton.

Hall, Stuart. 1992. "What Is This 'Black' in Black Popular Culture?" In *Black Popular Culture,* ed. Gina Dent, 21–36. Seattle, Wash.: Bay Press.

Hamon, Philippe. 1992. "The Major Features of Realist Discourse." In *Realism,.* ed. Lilian R. Furst, 166–85. London: Longman.

Handy, Bruce. 1997. "The Force is Back." *Time* 149, no. 6 (10 February): 68–74.

Hannerz, Ulf. 1970. "Another Look at Lower-Class Black Culture." In *Soul,* ed. Lee Rainwater, 167–86. Chicago: Aldine.

Hansen, Christian, Catherine Needham, and Bill Nichols. 1989. "Skin Flicks: Pornography, Ethnography, and the Discourses of Power." *Discourse* 11, no. 2:65–79.

Harris, Daniel. 1992. "Make My Rainy Day." *The Nation,* 8 June, 790–93.

Hartley, Howard W. 1925. *Tragedy of Sand Cave.* Louisville, Ky: Standard Printing.

Harvey, David. 1989. *The Condition of Postmodernity.* Oxford: Blackwell.

Haug, Wolfgang Fritz. 1986. *Critique of Commodity Aesthetics.* Cambridge: Polity Press.

Heidegger, Martin. [1962] 1977. "The Question Concerning Technology." In *The Question Concerning Technology and Other Essays,* trans. William Lovitt, 3–35. New York: Harper.

Heller, Terry. 1987. *The Delights of Terror: An Aesthetics of the Tale of Terror.* Urbana: University of Illinois Press.

Hendricks, Thomas. 1974. "Professional Wrestling as Moral Order." *Sociological Inquiry.* 44, no. 3:177–88.

Herf, Jeffery. 1984. *Reactionary Modernism: Technology, Culture, and Politics in the Weimar Republic and the Third Reich.* New York: Cambridge University Press.

Hershbell, Jackson B. 1978. "The Ancient Telegraph: War and Literacy." *Communication Arts in the Ancient World,* ed. Eric A. Havelock et al., 81–92. New York: Hastings.

Herskovits, Melville. 1941. *Myth of the Negro Past.* Boston: Beacon.

Heyer, Paul. 1988. *Communications and History: Theories of Media, Knowledge, and Civilization.* New York: Greenwood Press.

Hietala, Thomas R. 1985. *Manifest Design: Anxious Aggrandizement in Late Jacksonian America.* Ithaca: Cornell University Press.

Hirsch, Marianne. 1997. *Family Frames: Photography, Narrative, and Postmemory.* Cambridge: Harvard University Press.

Hitman Hart: Wrestling with Shadows. 1998. Film. Directed by Paul Jay. VHS. National Film Board of Canada.

Hoberman, J. 1985. "Spielbergism and Its Discontents." *Village Voice,* 9 July, 48+.

———. 1994. "Moondance." *Artforum International* 32, no. 10 (summer): 10+.

Hobsbawm, Eric. [1983] 1994. "The Nation as Invented Tradition." In Hutchinson and Smith 1994, 76–82.

Hoff, Joan. 1989. "Why Is There No History of Pornography?" In Gubar and Hoff 1989, 17–46.

Holbert, R. Lance. 1998. "A Critical Analysis of Marshall McLuhan's Radio-Fascism Probe." Paper presented at the Popular Culture Association Conference, Orlando, Fla.

Holbrook, Morris B. 1993. "On the New Nostalgia: 'These Foolish Things' and Echoes of the Dear Departed Past." In Browne and Ambrosetti 1993, 74–120.

Holmes, John Eric. 1981. *Fantasy Role Playing Games*. London and Melbourne: Arms and Armour Press.

The Howling. 1980. Film. Directed by Joe Dante. Script by Gary Brandner and John Sayles. Embassy.

hooks, bell. 1996a. *Cultural Criticism and Transformation*. Lecture on videotape. Northampton, Mass.: The Media Education Foundation.

———. 1996b. *Reel to Real: Race, Sex, and Class at the Movies*. New York: Routledge.

House Party I. 1990. Film. Directed by Reginald Hudlin. Screenplay by Reginald Hudlin. Produced by Warrington Hudlin. New Line.

House Party II. 1991. Film. Directed by George Jackson and Doug McHenry. Screenplay by Rusty Cundieff. New Line.

Houston, Beverle. 1984. "Viewing Television: The Metapsychology of Endless Consumption." *Quarterly Review of Film Studies* 9, no. 3 (summer): 183–95.

Howe, Andrew. 1999. "The Jerry Springer Show: Trash Television or the Revival of Old Comedy?" Session on Television Comedy, Popular Culture Association/American Culture Association Convention, San Diego Marriott Hotel, 3 April.

Hudlin, Warrington, and Reginald Hudlin. 1990. "They Gotta Have It." Interview with Marlaine Glicksman. *Film Comment* 26, no. 3:65–69.

Husserl, Edmund. [1928] 1950. *The Phenomenology of Internal Time-Consciousness*. Translated by J. S. Churchill. Bloomington: Indiana University Press.

———. 1969. *Formal and Transcendental Logic*. 1929. Reprint, The Hague: Martinus Nijhoff.

———. 1981. "Foundational Investigations of the Phenomenological Origin of the Spatiality of Nature." Translated by Fred Kersten. In *Husserl: Shorter Works*, ed. Peter McCormick and Frederick A. Elliston, 222–33. Notre Dame, Ind.: University of Notre Dame Press.

Hutchinson, John, and Anthony D. Smith, eds. 1994. *Nationalism.* New York: Oxford University Press.

Ice Cube. 1990. *AmeriKKKa's Most Wanted.* Compact disk. Priority Records.

―――. 1994. Interview, with bell hooks. In *Outlaw Culture: Resisting Representations,* by bell hooks, 125–43. London and New York: Routledge.

Ice Cube, and Dr. Dre. 1996. "Natural Born Killaz." Lyrics by Ice Cube and Dr. Dre. In *Murder Was the Case.* Snoop Doggy Dog, various artists. Compact disk. Priority Records.

Ihde, Don. 1971. *Hermeneutic Phenomenology: The Philosophy of Paul Ricoeur.* Evanston, Ill.: Northwestern University Press.

―――. 1976. *Listening and Voice: A Phenomenology of Sound.* Athens: Ohio University Press.

―――. 1979. *Technics and Praxis.* Dordrecht: D. Reidel.

―――. 1986. *Consequences of Phenomenology.* Albany: State University of New York Press.

―――. 1990. *Technology and the Lifeworld: From Garden to Earth.* Bloomington: Indiana University Press.

―――. 1993. *Postphenomenology: Essays in the Postmodern Context.* Evanston, Ill.: Northwestern University Press.

―――. 1996. "This Is Not a Text; or, Do We Read Images?" *Philosophy Today,* spring, 125–31.

IMDB (Internet Movie Data Base) 1999. http://us.imdb.com

Independence Day. 1996. Film. Directed by Roland Emmerlich. Screenplay by Dean Devlin. Twentieth Century Fox.

Innis, Harold A. 1951. *The Bias of Communication.* Toronto: University of Toronto Press.

Ivins, William M., Jr. 1973. *On the Rationalization of Sight.* New York: Da Capo.

Jackaway, Gwenyth L. 1995. *Media at War: Radio's Challenge to the Newspapers, 1924–1939.* Westport, Conn.: Praeger.

Jacobs, A. J. 1995. "Squeezing the Tube." *Entertainment Weekly,* 29 September, 8–9.

Jager, Bernd. 1985. "Body, House and City: The Intertwinings of Embodiment, Inhabitation and Civilization." In *Dwelling, Place, and Environment: Towards a Phenomenology of Person and World,* ed. David Seamon and Robert Mugerauer, 215–25. Dordrecht: Martinus Nijhoff.

Jameson, Fredric. 1989. "Nostalgia for the Present." *The South Atlantic Quarterly* 88, no. 2 (spring): 517–37.

Jay, Martin. 1988. "Scopic Regimes of Modernity." In *Vision and Visuality,* ed. Hal Foster, 3–28. Seattle, Wash.: Bay Press.

Jaynes, Julian. 1976. *The Origin of Consciousness in the Breakdown of the Bicameral Mind.* Boston: Houghton Mifflin.

Jenks, Chris. 1995a. "The Centrality of the Eye in Western Culture: An Introduction." In Jenks 1995b, 1–25.

———, ed. 1995b. *Visual Culture.* New York: Routledge.

Jenkins, Henry. 1997. "'Never Trust a Snake': WWF Wrestling as Masculine Melodrama." In *Out of Bounds: Sports, Media, and the Politics of Identity,* 48–77. Bloomington: Indiana University Press.

Johnson, Victoria E. 1993–94. "Polyphony and Cultural Expression: Interpreting Musical Traditions in *Do the Right Thing.*" *Film Quarterly* 47, no. 2 (winter): 18–29.

Jones, Edgar R. 1979. *Those Were the Good Old Days: A Happy Look at American Advertising, 1880–1950.* New York: Simon & Schuster.

Jones, Ernst. 1931. *On the Nightmare.* London: Hogarth Press.

Jones, Quincy. 1995. Interview. In *Up From the Underground,* vol. 11 of *The History of Rock and Roll.* Time/Life Videotape.

Jones, Steve. 1992. *Rock Formation: Music, Technology, and Mass Communication.* Newberry Park, Calif.: Sage.

Joyce, James. [1922] 1961. *Ulysses.* New York: Modern Library.

Juice. 1992. Directed by Ernst R. Dickerson. Paramount.

Jünger, Ernst. 1929. *Storm of Steel: From the Diary of a German Storm-Troop Officer on the Western Front.* Trans. Basil Creighton. New York: Doubleday.

Kafka, Franz. 1983. "A Hunger Artist." In *Kafka: The Complete Stories and Parables,* trans. Willa Muir and Edwin Muir. New York: QPB. First published in *Ein Hungerkunstler: Vier Geschichten* (Berlin: Verlag Die Schmiede, 1924).

Kaha, C. W. 1994. "Of Boundaries and Variable-Flex Space." *Proteus* 11, no. 1 (spring): 18–20.

Kant, Immanuel. [1790] 1914. *Critique of Pure Judgement.* Trans. J. H. Bernard. 2d ed., rev. London: Macmillan.

Kaplan, E. Ann. 1987. *Rocking Around the Clock.* New York: Routledge.

Katz, David. 1950. *Gestalt Psychology.* Trans. Robert Tyson. New York: Ronald Press.

Keil, Charles. 1966. *Urban Blues.* Chicago: University of Chicago Press.

Kellerman, Ahron. 1989. *Time, Space, and Society: Geographical Societal Perspectives.* Dordrecht: Kluwer Academic.

Kellner, Douglas. 1992. *The Persian Gulf TV War.* Boulder, Colo.: Westview.

Kelly, Robin. 1994. *Race Rebels*. New York: Free Press.

Kendrick, Walter. 1987. *The Secret Museum: Pornography in Modern Culture*. New York: Viking.

Kermode, Frank. 1975. Introduction to *Selected Prose of T. S. Eliot*, ed. Frank Kermode. New York: Harcourt Brace Jovanovich and Farrar, Straus, and Giroux.

Kern, Stephen. 1976. *Anatomy and Destiny: A Cultural History of the Human Body*. Indianapolis, Ind.: Bobbs-Merrill.

———. 1983. *The Culture of Time and Space*. Harvard University Press.

———. 1991. "Wireless World." In Crowley and Heyer 1991, 186–89.

Kittler, Frederick A. 1990. *Discourse Networks 1800/1900*. Palo Alto, Calif.: Stanford University Press.

Kimmel, Michael S. 1990a. "'Insult' or 'Injury': Sex, Pornography, and Sexism." In Kimmel 1990b, 305–19.

———, ed. 1990b. *Men Confront Pornography*. New York: Crown Press.

Kinder, Marsha. 1991. *Playing with Power in Movies, Television, and Video Games: From Muppet Babies to Teenage Mutant Ninja Turtles*. Berkeley: University of California Press.

King, Geoff. 1996. *Mapping Reality: A Exploration of Cultural Cartographies*. New York: St. Martin's Press.

Kinsey Institute, Indiana University, Bloomington, Ind. Pornographic Photography Collection: Ref. no. 34043. Circa 1934. Possible "cum shot." Ref. no. 5306. Circa 1915. Priest blessing masturbating nuns. Ref. no. 5401. Circa 1915. Altar boy sodomizing a priest. Cf. nos. 5301–63 and 5401–53.

Kirby, Doug, Ken Smith, and Mike Wilkins. 1996–97. "Roadside America: Floyd Collins Museum." ONLINE: www.roadsideamerica.com/attract/KYCAVfloyd.html

Klein, Kerwin Lee. 1997. *Frontiers of the Historical Imagination*. Berkeley: University of California Press.

Klein, Sali J. 1992. *The Degeneration of Women: Bram Stoker's Dracula as Allegorical Criticism of the Fin de Siècle*. Rheinbach-Merzbach: CMZ-Verlag.

Klemm, David E. 1983. *The Hermeneutical Theory of Paul Ricoeur*. Lewisburg, Pa.: Bucknell University Press.

Kockelmans, Joseph J. 1967. *Phenomenology: The Philosophy of Edmund Husserl and Its Interpretations*. Garden City, N.Y.: Doubleday.

Kristeva, Julia. 1982. *The Powers of Horror*. New York: Columbia University Press.

Kroll, Jack. 1999. "Porn o' Plenty: XXX Fare Is Here." *Newsweek*, 5 April, 70.

Kundera, Milan. 1980. *The Book of Laughter and Forgetting.* Trans. Michael Henry Heim. New York: Knopf.

Lasch, Christopher. 1977. *Haven in a Heartless World.* New York: Norton.

———. 1990. "Memory and Nostalgia, Gratitude and Pathos." *Salmagundi,* winter–spring, 18–25.

Lauter, Paul. 1983. "Race and Gender in the Shaping of the American Canon: A Case Study from the Twenties." *Feminist Studies* 9, no. 3:435–63.

Lawrence, Amy. 1991. *Echo and Narcissus: Women's Voices in Classical Hollywood Cinema.* Berkeley: University of California Press.

Leab, Daniel J. 1975. *From Sambo to Superspade: The Black Experience in Motion Pictures.* Boston: Houghton Mifflin.

Lears, T. J. Jackson. 1983. "From Salvation to Self-Realization: Advertising and the Therapeutic Roots of the Consumer Culture, 1880–1930." In *The Culture of Consumption: Critical Essays in American History, 1880–1980,* ed. Richard Wightman Fox and T .J. Jackson Lears, 1–38. New York: Pantheon.

Legman, Gershon. 1975. *Rationale of the Dirty Joke: An Analysis of Sexual Humor.* 2d ser. New York: Bell Publishing Co.

Leland, John. 1998. "Stone Cold Crazy." *Newsweek,* 23 November, 58–64.

Levidow, Les. 1995. "The Gulf War: Castrating the Other." *Psychoculture* 1, no. 1 (spring): 9–15.

Levine, Lawrence. 1988. *Highbrow/Lowbrow: The Emergence of Cultural Hierarchy in America.* Cambridge: Harvard University Press.

Lewis, George H. 1993. "Bringing It All Back Home: Uses of the Past in the Present (and the Future) of American Popular Music." In Browne and Ambrosetti 1993, 61–73.

Lipsitz, George. 1990. *Time Passages: Collective Memory and Popular Culture.* Minneapolis: University of Minnesota Press.

———. 1995. "The Possessive Investment in Whiteness: Racialized Social Democracy and the 'White' Problem in American Studies." *American Quarterly* 47, no. 3:369–78.

Locke, Alain. 1925. "The Legacy of the Ancestral Arts." In *The New Negro: An Interpretation,* ed. Alain Locke, 254–67. New York: Alfred and Charles Boni.

London, Bette. 1990. *The Appropriated Voice: Narrative Authority in Conrad, Forster, and Woolf.* Ann Arbor: University of Michigan Press.

Long, Edward V. 1967. *The Intruders.* New York: Praeger.

Lorentzen, Justin J. 1995. "Reich Dreams: Ritual Horror and Armoured Bodies." In Jenks 1995, 161–69.

Lovecraft, H. P. [1920] 1971a. "From Beyond." In Lovecraft 1971d, 59–66.

———. [1925] 1971b. "The Lurking Fear" In Lovecraft 1971d, 1–22.

———. [1936] 1971c. "The Shadow Over Innsmouth." In Lovecroft 1971d, 115–82.

———. 1971d. *The Lurking Fear and Other Stories*. New York: Ballantine.

Lovett-Graff, Bennett. 1997. "Shadows over Lovecraft: Reactionary Fantasy and Immigrant Eugenics." *Extrapolation: A Journal of Science Fiction and Fantasy* 38, no. 3:175–92.

Lubar, Steven. 1993. *InfoCulture*. Boston: Houghton Mifflin.

Ludington, Townsend. 1980. *John Dos Passos: A Twentieth-Century Odyssey*. New York: Dutton.

Lull, James. 1991. "Understanding Radio." In Crowley and Heyer 1991, 207–12.

———. 1995. *Media, Communication, and Culture*. Cambridge: Blackwell.

Lurie, Alison. 1990. *Don't Tell the Grown-Ups: Why Kids Love the Books They Do*. New York: Avon.

MacDonald, Dwight. 1957. "A Theory of Mass Culture." In *Mass Culture: The Popular Arts in America,* ed. Bernard Rosenberg and David Manning White, 59–73. Glencoe, Ill.: Free Press.

Madsen, Axel. 1969. *Billy Wilder*. Bloomington: Indiana University Press.

Malinowski, Bronislaw. 1944. *Freedom and Civilization*. New York: Roy.

———. 1946. *Dynamics of Culture Change*. Glenwood, Ill.: Greenwood.

———. 1954. *Magic, Science, and Religion*. Glenwood, Ill.: Greenwood.

Maraval, José Antonio. 1986. *Culture of the Baroque*. Minneapolis: University of Minnesota Press,.

Marc, David. *Demographic Vistas*. 1984. Rev. ed. Philadelphia: University of Pennsylvania Press.

Marchand, Roland. 1985. *Advertising the American Dream: Making Way for Modernity, 1920–1940*. Berkeley: University of California Press.

Marcus, Stephen. 1964. *The Other Victorians*. London: Weidenfeld and Nicolson.

Marinetti, Filippo Tommaso. 1972. *Selected Writings*. Edited by R. W. Flint. New York: Farrar, Straus, and Giroux.

Martin, Michelle. 1991. *"Hello, Central?": Gender, Technology, and Culture in the Formation of Telephone Systems*. Montreal: McGill-Queens University Press.

Marvin, Carolyn. 1988. *When Old Technologies Were New: Thinking About Electric Communication in the Late Nineteenth Century*. New York: Oxford University Press.

Marx, Karl. [1867] 1978. *Das Kapital*. In *The Marx-Engels Reader,* ed. Robert C. Tucker. New York: W. W. Norton.

Masilela, Ntongela. 1993. "The Los Angeles School of Black Filmmakers." In Diawara 1993b, 107–17.

Maxwell, William. 1991. "Sampling Authenticity: Rap Music, Postmodernism, and the Ideology of Black Crime." *Studies in Popular Culture* 14, no. 1:1–15.

May, Lary. 1980. *Screening Out the Past: The Birth of Mass Culture and the Motion Picture Industry*. New York: Oxford University Press.

McClain, Leanta. 1980. "The Black Middle-Class Burden." *Newsweek*, 13 October, 21.

McGuigan, J. 1992. *Cultural Populism*. London: Routledge.

McLuhan, Marshall. 1964. *Understanding Media*. 2d ed. New York: Signet.

McNeill, William H. 1986. *Mythhistory and Other Essays*. Chicago: University of Chicago Press.

Menace II Society. 1993. Film. Directed by Alan Hughes and Albert Hughes. Screenplay by Alan Hughes, Albert Hughes, and Tyger Williams. New Line.

"Metal Up Your Ass: The Why Rap Sucks Page." 1998. http://php.iupui.edu/~jjgoins/rap/petition.html

Metcalf, Robert M. 1994. "Maybe We Shouldn't Be Calling It Cyberspace." *Infoworld* 16, no. 17 (25 April): 63.

Meyrowitz, Joshua. 1985. *No Sense of Place: The Impact of Electronic Media on Social Behavior*. New York: Oxford University Press.

Michaelis, Meir. 1978. *Mussolini and the Jews*. New York: Oxford University Press.

Michelson, Peter. 1971. *The Aesthetics of Pornography*. New York: Herder & Herder.

———. 1993. *Speaking the Unspeakable: A Poetics of Obscenity*. Albany: State University of New York Press.

Miles, Margaret R. 1985. *Image as Insight: Visual Understanding in Western Christianity and Secular Culture*. Boston: Beacon.

Miller, George J., ed. 1937. *Activities in Geography*. Bloomington, Ill.: McKnight & McKnight.

Miller, Izchak. 1984. *Husserl, Perception, and Temporal Awareness*. Cambridge: MIT Press.

Miller Frank, Felicia. 1995. *The Mechanical Song: Women, Voice, and the Artificial in Nineteenth-Century French Narrative*. Palo Alto, Calif.: Stanford University Press.

Mintz, Sidney W. 1985. *Sweetness and Power: The Place of Sugar in Modern History*. New York: Viking.

Mintz, Sidney W., and Richard Price. 1976. *The Birth of African American Culture*. Philadelphia: Institute for Human Issues.

Mitchell, Breon. "Kafka and the Hunger Artists." In *Kafka and the Contemporary Critical Performance*, ed. Alan Udoff, 236–55. Bloomington: Indiana University Press.

Mitchell, W. J. T. 1986. *Iconology: Image, Text, Ideology*. Chicago: University of Chicago Press.

Molesworth, Charles. 1986. "Rambo, Passion, and Power." *Dissent* 33 (winter): 109–11.

Molz, R. Kathleen. 1978. "Reality and Reason: Intellectual Freedom and Youth." In *Young Adult Literature in the Seventies.*, ed. Jana Varlejs, 200–202. Metuchen, N.J.: Scarecrow Press.

Monmonier, Mark. 1995. *Drawing the Line: Tales of Maps and Cartocontroversy*. New York: Henry Holt and Co.

Moog, Carol. 1990. *Are They Selling Her Lips?* New York: William Morrow.

Morais, Vamberto. 1976. *A Short History of Anti-Semitism*. New York: Norton.

Morley, David. 1995. "Television: Not So Much a Visual Medium, More a Visible Object." In Jenks 1995, 170–89.

Morris, Edmund. 1999. *Dutch: A Memoir of Ronald Reagan.* New York: Random House.

Morrison, Toni. 1977. *Song of Solomon*. New York: New American Library.

Morton, Gerald W., and George O'Brien. 1985. *Wrestling to Rasslin: Ancient Sport to American Spectacle*. Bowling Green, Ohio: Bowling Green State University Press.

Moses, Wilson Jeremiah. 1998. *Afrotopia: The Roots of African American Popular History*. Cambridge and New York: Cambridge University Press.

Mosse, George L. 1985. *Nationalism and Sexuality*. New York: H. Fertig.

———. 1996a. "Fascist Aesthetics and Society: Some Considerations" *Journal of Contemporary History* 31, no. 2 (April): 245–52. Special issue: "The Aesthetics of Fascism."

———. 1996b. *The Image of Man: The Creation of Modern Masculinity*. New York: Oxford University Press.

Mumford, Lewis. [1964] 1970. *The Myth of the Machine: The Pentagon of Power*. New York: Harcourt Brace Jovanovich.

Murray, Robert K., and Roger W. Brucker. 1979. *Trapped! The Story of the Struggle to Rescue Floyd Collins*. New York: Putnam.

Nadel, Alan. 1997. *Flatlining on the Field of Dreams: Cultural Narratives in the Films of President Reagan's America*. New Brunswick, N.J.: Rutgers University Press.

Nelson, Havelock, and Michael A. Gonzales. 1991. *Bring the Noise: A Guide to Rap Music and Hip-Hop Culture.* New York: Harmony.

Nelson, Jenny L. 1990. "The Dislocation of Time: A Phenomenology of Television Reruns." *Quarterly Review of Film and Video* 12, no. 3:79–92.

Neuman, W. Russell. 1991. *The Future of the Mass Audience.* Cambridge: Cambridge University Press.

Nietzsche, Friedrich. 1954. "On Truth and Lie in an Extra Moral Sense." In *The Portable Nietzsche,* ed. Walter Kaufmann, 42–50. New York: Penguin.

———. [1872] 1967. *The Birth of Tragedy and The Case of Wagner.* Trans. Walter Kaufmann. New York: Vintage.

Oliver, Melvin L., and Thomas M. Shapiro. 1995. *Black Wealth/White Wealth: A New Perspective on Racial Inequality.* New York: Routledge.

O'Meally, Robert G. 1980. *The Craft of Ralph Ellison.* Cambridge: Harvard University Press.

Ong, Walter J. 1977. *Interfaces of the Word.* Ithaca: Cornell University Press.

Oston, Mortimer. 1996. *Myth and Madness: The Psychodynamics of Anti-Semitism.* New Brunswick, N.J.: Transaction.

Otto, Rudolf. 1928. *The Idea of the Holy.* Trans. John W. Harvey. London: Oxford University Press.

Ozersky, Josh. 1991. "TV's Anti-Families: Married . . . with Malaise." *Tikkun* 6, no. 1 (January/February): 11.

Paglia, Camille. 1990. *Sexual Personae.* New York: Vintage.

Pardun, Carol J., and Kathy B. McKee. 1995. "Strange Bedfellows: Symbols of Religion and Sexuality on MTV." *Youth and Society* 26, no. 4 (June): 438–49.

Patnode, Randall. 1998. "Everyman's Ether: The Social Construction of Broadcasting by Two New York Newspapers, 1922–1924." Paper presented at the Popular Culture Association Convention. Orlando, Fla., April.

Payne, Stanley. 1995. *A History of Fascism: 1914-1945.* Madison: University of Wisconsin Press.

Pecora, Vincent. 1985. "*Heart of Darkness* and the Phenomenology of Voice." *English Literary History* 52, no. 4 (winter): 993–1015.

Pennebaker, D. A., dir. [1968] 1986. *Jimi Hendrix Live at Monterey.* Videodisk. HBO Video.

Pepsi Generation Collection. 1984. Completed in 1984 for the Modern Advertising History Projects, Archives Center, National Museum of American History, Smithsonian Institution, Washington, D.C.; a collection of advertising artifacts, documents, and interviews on audiotape compiled by oral historian Scott Ellsworth. Specific Documents consulted: Project

Briefing Book; Walter Mack, president, Pepsi-Cola,and other BBD&O and Pepsi executives and creative staff (Pottash, Dillon, Ramin, Durkee, Dusenberry, Rosenshire, Lipsitz, et al.

Pepsi-Cola Television Commercials. 1985. Videotape. "Guitar." AT 86: 0518. The Museum of Television and Radio, Beverly Hills, California.

Pepsi-Cola Television Commercials. 1988. Videotape. With Cindy Crawford. AT 28905. The Museum of Television and Radio, Beverly Hills, California.

Perkins, William Eric. 1996a. "The Rap Attack: An Introduction." In Perkins 1996b, 1–45.

———, ed. 1996b. *Droppin' Science: Critical Essays on Rap Music and Hip Hop Culture*. Philadelphia: Temple University Press.

Pettegrew, John. "A Post-Modernist Moment: 1980s Commercial Culture and the Founding of MTV." *Journal of American Culture* 15, no. 4 (winter): 57–65.

Pickles, John. 1985. *Phenomenology, Science and Geography: Spatiality and the Human Sciences*. Cambridge and New York: Cambridge University Press.

Pinkus, Karen. 1995. *Bodily Regimes: Italian Advertising under Fascism*. Minneapolis: University of Minnesota Press.

Polkinghorne, D. E. 1988. *Narrative Knowing and the Human Sciences*. Albany: State University of New York Press.

Pollari, Pat. 1996. *The Great Puke-Off*. New York: Bantam.

Pratt, Linda Ray. 1994. "Liberal Education and the Idea of the Postmodern University." *Academe* 80, no. 6 (November/December): 46–51.

Puttnam, David. 1998. *Movies and Money*. New York: Knopf.

Rainwater, Lee, ed. 1970. *Soul*. Chicago: Aldine.

Redman, Tim. 1991. *Ezra Pound and Italian Fascism*. Cambridge and New York: Cambridge University Press.

Retrosex. Com: Vintage Erotica. 1999. www.retrosex.com.

Reynolds, Simon, and Joy Press. 1995. *The Sex Revolts: Gender, Rebellion, and Rock 'n' Roll*. Cambridge: Harvard University Press.

Richards, T. 1991. *The Commodity Culture of Victorian England: Advertising and the Spectacle, 1851–1914*. London: Verso.

Ricoeur, Paul. 1967. *The Symbolism of Evil*. New York: Harper and Row.

———. 1974. *The Conflict of Interpretations: Essays in Hermeneutics*. Ed. Don Ihde. Evanston, Ill.: Northwestern University Press.

———. 1984–85. *Time and Narrative*. Vols. 1–3. Chicago: University of Chicago Press.

Robins, Kevin. 1996. *Into the Image: Culture and Politics in the Field of Vision.* London and New York: Routledge.

Rogin, Michael. 1998. *Independence Day, or How I Learned to Stop Worrying and Love the Enola Gay.* London: British Film Institute.

Robocop. 1987. Film. Directed by Paul Verhoeven. Script by Michael Miner and Edward Neumeier. Orion.

Robocop File. American Film Institute Collection at the University of Southern California, Los Angeles, Calif..

Roemer, Michael. 1995. *Telling Stories: Postmodernism and the Invalidation of Traditional Narrative.* Lanham, Md.: Rowman & Littlefield.

Roosevelt, Franklin D. 1938a. "The First 'Fireside Chat'—An Intimate Talk with the People of the United States on Banking. March 12, 1933." In Roosevelt 1938c, 61–65.

———. 1938b. "Inaugural Address. March 4, 1933." In Roosevelt 1938c, 11–16.

———. 1938c. *The Public Papers and Addresses of Franklin D. Roosevelt.* Vol. II: *The Year of Crisis, 1933.* Ed. Samuel I. Rosenman. New York: Random House.

Rose, Tricia. 1994. *Black Noise: Rap Music and Black Culture in Contemporary America.* Hanover, N.H.: University Press of New England.

Ross, Andrew. 1989. *No Respect: Intellectuals and Popular Culture.* New York: Routledge.

Ross, Steven J. 1998. *Working-Class Hollywood: Silent Film and the Shaping of Class in America.* Princeton: Princeton University Press.

Ross, Stephen M. 1989. *Fiction's Inexhaustible Voice: Speech and Writing in Faulkner.* Athens: University of Georgia Press.

Roszak, T. 1968. "The Summa Popologia of Marshall McLuhan." In *McLuhan: Pro and Con,* ed. R. Rosenthal, 257–69. Baltimore: Penguin, 1968.

Rubin, Theodore Isaac. 1990. *Anti-Semitism: A Disease of the Mind.* New York: Continuum.

Ryan, Halford R. 1988. *Franklin D. Roosevelt's Rhetorical Presidency.* New York: Greenwood Press.

Said, Edward. 1978. *Orientalism.* New York: Pantheon.

Salvaggio, Jerry L. 1987. "Projecting a Positive Image of the Information Society." In *The Ideology of the Information Age,* ed. Jennifer Daryl Slack, and Fred Fejes, 146–57. Norwood, N.J.: Ablex.

Schiller, Herbert I. 1969. *Mass Communications and American Empire.* New York: Kelly.

Scholes, Robert, and Eric S. Rabkin. 1977. *Science Fiction: History, Science, Vision.* New York: Oxford University Press.

Scholes, Robert, and Robert Kellog. 1966. *The Nature of Narrative.* Oxford: Oxford University Press.

Schudson, Michael. 1978. *Discovering the News.* New York: Basic Books.

Schusterman, Richard. 1991. "The Fine Art of Rap." *New Literary History* 22:613–32.

Schutz, Alfred. 1962. "On Multiple Realities." In *Collected Papers, Vol. 1,* ed. Maurice Nathanson, 207–86. The Hague: Martinus Nijhoff.

Schutz, Alfred, and Thomas Luckman. 1973. *The Structures of the Life-World.* Evanston, Ill.: Northwestern University Press.

Seelye, John. 1998. *Memory's Nation: The Place of Plymouth Rock.* Chapel Hill: University of North Carolina Press.

Seeßlen, Georg. 1994. *Tanz den Adolf Hitler: Faschismus in der populären Kultur.* Berlin: Edition Tiamat.

Seton-Watson, Hugh. [1977] 1994. "Old and New Nations." In *Nationalism,* ed. John Hutchinson and Anthony D. Smith. Oxford and New York: Oxford University Press.

Shakur, Tupac. 1995. *Me Against the World.* Interscope Records.

Shames, Lawrence. 1989. *The Hunger for More: Searching for Values in an Age of Greed.* New York: Random House.

Shaw, Harry, ed. 1990. *Perspectives of Black Popular Culture.* Bowling Green, Ohio: Popular Press.

Silverman, Kaja. 1988. *The Acoustic Mirror.* Bloomington: Indiana University Press.

———. 1996. *The Threshold of the Visible World.* New York and London: Routledge.

Sing, Bing, Sing. 1931. Film. Produced by Mack Sennett. Paramount.

Singin' in the Rain. 1952. Film. Directed by Stanley Donen and Gene Kelly. Screenplay by Betty Comden and Adolph Green.. MGM.

Slade, Joseph W. 1997. "Pornography in the Late Nineties." *Wide Angle* 19, no. 3 (July): 1–12.

Slater, Don. 1995. "Photography and Modern Vision: The Spectacle of 'Natural Magic.'" In Jenks 1995b, 218–37.

Slotkin, Richard. 1973. *Regeneration through Violence: The Mythology of the American Frontier, 1600–1860.* Middletown, Conn.: Wesleyan University Press.

———. 1992. *Gunfighter Nation.* New York: HarperCollins.

Snyder, Ross. 1979. "Architects of Contemporary Man's Consciousness." In *Inter/Media,* ed. Gary Gumpert and Robert Cathcart, 350–60. New York: Oxford University Press.

Soja, Edward. 1989. *Postmodern Geographies: The Reassertion of Space in Critical Social Theory*. London: Verso.

Sola Pool, Ithiel de. 1977. *The Social Impact of the Telephone*. Cambridge: MIT Press.

———. 1990. *Technologies without Boundaries: On Telecommunications in a Global Age*. Cambridge: Harvard University Press.

Sontag, Susan. 1973. *On Photography*. New York: Farrar, Straus and Giroux.

Spigel, Lynn. 1990. "The Domestic Gaze." In *From Receiver to Remote Control: The TV Set,* ed. Matthew Geller, 11–22. New York: New Museum of Contemporary Art.

Stallybrass, Peter, and Allon White. 1986. *The Politics and Poetics of Transgression*. Ithaca: Cornell University Press.

Stambovsky, Phillip. 1988. *The Depictive Image*. Amherst: University of Massachusetts Press.

Star Trek: Generations. 1994. Film. Directed by David Carson. Screenplay by Ronald Moore and Brannon Braga. Paramount.

"Star Wars Nostalgia." 1997. On *Access Hollywood* (television program), 9 February.

Steele, Shelby. 1988. "On Being Black and Middle Class." *Commentary*, January, 42–47.

Sternhell, Zeev. 1976. "Fascist Ideology." In *Fascism: A Reader's Guide*, ed. Walter Laqueur, 315–76. Berkeley: University of California Press.

———. 1995. *Neither Right Nor Left: Fascist Ideology in France*. Princeton: Princeton University Press.

Stoker, Bram. [1899] 1978. *Dracula*. Reprinted in *Frankenstein, Dracula, and Dr. Jekyll and Mr. Hyde*. New York: Signet.

Stone, Allucquere Rosanne. 1995. *The War of Desire and Technology at the Close of the Mechanical Age*. Cambridge: MIT Press.

Strauss, Bob. 1995. "What About 'Bob?'" *Entertainment Weekly,* 14 April, 67–68.

Taussig, H. Arthur. 1990. *Robocop—The Truly Modern Prometheus*. Costa Mesa, Calif.: H. Artur Taussig.

Taylor, Mark C., and Esa Saarinen. 1994. *Imagologies: Media Philosophy*. New York: Routledge.

Taxi Driver. 1976. Film. Directed by Martin Scorsese. Columbia.

Theweleit, Klaus. 1987. *Male Fantasies*. Vol. 1. Minneapolis: University of Minnesota Press.

Thomas, Tony. 1980. *The Films of Ronald Reagan*. Secaucus, N.J.: Citadel Press.

Thompson, Daniel C. 1986. *A Black Elite.* Westport, Conn: Greenwood Press.

Thompson, Robert Farris. 1983. *Flash of the Spirit: African and Afro-American Art and Philosophy.* New York: Random House.

Tierney, John. 1994. "Pornography, the Low-Slung Engine of Progress." *The New York Times,* 9 January, sec 2.1.18.

Tomasulo, Frank P. 1994. "The Spectator in the Tube: The Rhetoric of *Donahue.*" *Journal of Film and Video* 36 (spring): 5–12.

Toop, David. 1984. *The Rap Attack: African Jive to New York Hip Hop.* Boston: South End Press.

Townsend, Pete. 1995. Interview. In *Guitar Heroes,* vol. 10 of *The History of Rock and Roll.* Time/Life Videotape.

Tropic Zone. 1953. Film. Directed by Lewis R. Foster. Screenplay by Lewis R. Foster. Paramount. Based on the novel *Gentleman of the Jungle,* by Tom Gill (New York: Dell, 1940).

Turkle, Sherry. 1995. *Life on the Screen: Identity in the Age of Internet.* New York: Simon & Schuster.

Turnbull, David, and Helen Watson. 1993. *Maps Are Territories.* Chicago: University of Chicago Press.

Turner, Frederick Jackson. 1962. *The Frontier in American History.* Foreword by Ray Allen Billington. New York: Holt, Rinehart, and Winston.

Turner, Richard. 1996. "An Ear for the CBS Eye." *Newsweek* 128, no. 25 (16 December): 58–59.

Twitchell, James. 1985. *Dreadful Pleasures: An Anatomy of Modern Horror.* New York: Oxford University Press.

———. 1989. *Preposterous Violence.* New York: Oxford University Press.

Tye, Larry. 1998. *The Father of Spin: Edward L. Bernays and the Birth of Public Relations.* New York: Crown.

Vaughn, Stephen. 1994. *Ronald Reagan in Hollywood: Movies and Politics.* New York: Cambridge University Press.

Verhoeven, Paul. 1987. Interview. *American Film,* October, 33–35.

Voller, Jack G. 1993. "Neuromanticism: Cyberspace and the Sublime." *Extrapolation* 34, no. 1:18–29.

Walker, Alice. 1990. "Everyday Use." In *The Norton Anthology of Short Fiction,* 4th ed., ed. R. V. Cassill, 1631–39. New York: Norton, 1990.

Wallace, Michelle. 1990. "Modernism, Postmodernism and the Problem of the Visual in Afro-American Culture" In *Out There: Marginalization and Contemporary Cultures,* ed. Russell Ferguson et al., 39–50. Cambridge: MIT Press.

Walser, Robert. 1993. *Running with the Devil: Power, Gender, and Madness in Heavy Metal Music.* Middletown, Conn.: Wesleyan University Press.

Walton, Kendall L. 1978. "How Remote Are Fictional Worlds from the Real World?" *Journal of Aesthetics and Art Criticism* 37, no. 1:12–23.

Warren, Donald. 1996. *Radio Priest.* New York: Free Press.

Watt, Ian. 1964. *The Rise of the Novel.* Berkeley: University of California Press.

Weaver, Mary Jo. 1989. "Pornography and the Religious Imagination." In Gubar and Hoff 1989, 68–86.

Welsh, David. 1993. *The Third Reich: Politics and Propaganda.* London and New York: Routledge.

Wernick, Andrew. 1991. *Promotional Culture: Advertising Ideology and Symbolic Expression.* London and Newbury Park, Calif.: Sage.

West, Cornel. 1993. *Race Matters.* Boston: Beacon.

Westling, Donald, and Tadeusz Slawek. 1995. *Literary Voice: The Calling of Jonah.* Albany: State University of New York Press.

White, Hayden. 1981. "The Value of Narrativity in the Representation of Reality." In *On Narrative*, ed. W. J. T. Mitchell, 1–24. Chicago: University of Chicago Press.

White, Shane, and Graham White. 1998. *Stylin': African American Expressive Culture from Its Beginnings to the Zoot Suit.* Ithaca and London: Cornell University Press.

Williams, Linda. 1989. "Fetishism and Hard Core: Marx, Freud, and the 'Money Shot.'" In Gubar and Hoff 1989, 198–217.

———. 1999. *Hard Core: Power, Pleasure, and the Frenzy of the Visible.* Rev. ed. Berkeley: University of California Press.

Williams, Raymond. 1975. *Television: Technology and Cultural Form.* New York: Schocken.

———. [1980] 1993. "Advertising: The Magic System." In *The Cultural Studies Reader*, ed. Simon During, 320–36. London and New York: Routledge.

Wills, Gary. 1988. *Reagan's America.* New York: Penguin.

Wilson, Rob. 1994. "Cyborg America: Policing the Social Sublime in *Robocop* and *Robocop 2.*" In *The Administration of Aesthetics: Censorship, Political Critique, and the Public Sphere*, ed. Richard Burt, 289–306. Minneapolis, University of Minnesota Press.

Winrock, Michael. 1990. *Nationalisme, antisemitisme, et fascisme.* Paris: Seuil.

Wood, Denis (with John Fels). 1992. *The Power of Maps.* New York and London: Guilford Press.

Wurtzel, Alan H., and Colin Turner. 1977. "Latent Functions of the Telephone." In Sola Pool 1977, 246–61.

Yack, Bernard. 1997. *The Fetishism of Modernities.* Notre Dame, Ind.: University of Notre Dame Press.

Young, Katherine G. 1987. *Taleworlds and Storyrealms.* Dordrecht: Martinus Nijhoff.

ZZ Top's Greatest Hits. 1992. Compact disk. Warner Brothers.

Index

231